Instructor's Guide and Solutions Manual to Organic Structures from 2D NMR Spectra

Instructor's Guide and Solutions Manual to Organic Structures from 2D NMR Spectra

L. D. Field, H. L. Li and A. M. Magill
School of Chemistry, University of New South Wales, Australia

Library of Congress Cataloging-in-Publication Data applied for.

A catalogue record for this book is available from the British Library.

ISBN: 9781119027256

Set in 12/18pt Times New Roman by Aptara Inc., New Delhi, India.

Printed and bound by CPI Group (UK) Ltd, Croydon CR0 4YY

1   2015

CONTENTS

Contents

PREFACE

This book is the Instructor's Guide and Solutions Manual to the problems contained in the text *Organic Structures from 2D NMR Spectra*.

The aim of this book is to teach students to solve structural problems in organic chemistry using NMR spectroscopy and in particular 2D NMR spectroscopy. The basic philosophy of the book is that learning to identify organic structures using spectroscopy is best done by working through examples. This book contains a series of about 60 graded examples ranging from very elementary problems through to very challenging problems at the end of the collection.

We have assumed a working knowledge of basic structural organic chemistry and common functional groups. We also assume a working knowledge of the rudimentary spectroscopic methods which would be applied routinely in characterising and identifying organic compounds including infrared spectroscopy and basic 1D ^{13}C and ^{1}H NMR spectroscopy.

The Instructor's Guide contains a worked solution to each of the problems contained in *Organic Structures from 2D NMR Spectra*. At the outset, it should be emphasised that there are always many paths to the correct answer – there is no single process to arrive at the correct solution to any of the problems. We do not recommend a mechanical attitude to problem solving – intuition, which comes with experience, has a very important place in solving structures from spectra; however, students often find the following approach useful:

(i) **Extract as much information as possible from the basic characterisation data which is provided:**

   (a) **Note the molecular formula** and any restrictions this places on the functional groups that may be contained in the molecule.

   (b) From the molecular formula, **determine the degree of unsaturation**. The degree of unsaturation can be calculated from the molecular formula for all compounds containing C, H, N, O, S and the halogens using the following three basic steps:

      1. Take the molecular formula and replace all halogens by hydrogens.

2. Omit all of the sulfur and/or oxygen atoms.

3. For each nitrogen, omit the nitrogen and omit one hydrogen.

After these three steps, the molecular formula is reduced to C_nH_m, and the degree of unsaturation is given by:

$$\text{Degree of Unsaturation} = n - {}^m/_2 + 1$$

The degree of unsaturation indicates the number of π bonds and/or rings that the compound contains. For example, if the degree of unsaturation is 1, the molecule can only contain one double bond or one ring. If the degree of unsaturation is 4, the molecule must contain four rings or multiple bonds. An aromatic ring accounts for four degrees of unsaturation (the equivalent of three double bonds and a ring). An alkyne or a C≡N accounts for two degrees of unsaturation (the equivalent of two π bonds).

(c) **Analyse the 1D ^{1}H NMR spectrum** if one is provided and note the relative numbers of protons in different environments and any obvious information contained in the coupling patterns. Note the presence of aromatic protons, exchangeable protons, and/or vinylic protons, all of which provide valuable information on the functional groups which may be present.

(d) **Analyse the 1D ^{13}C NMR spectrum** if one is provided and note the number of carbons in different environments. Note also any resonances that would be characteristic of specific functional groups, *e.g.* the presence or absence of a ketone, aldehyde, ester or carboxylic acid carbonyl resonance.

(e) **Analyse any infrared data** and note whether there are absorptions characteristic of specific functional groups, *e.g.* C=O or –OH groups.

(ii) **Extract basic information from the 2D COSY, TOCSY and/or C–H correlation spectra.**

(a) The COSY will provide obvious coupling partners. If there is one identifiable starting point in a spin system, the COSY will allow the successive identification (*i.e.* the sequence) of all nuclei in the spin system. The COSY cannot jump across breaks in the spin system (such as where there is a heteroatom or a carbonyl group that isolates one spin system from another).

(b) The TOCSY identifies all groups of protons that are in the same spin system.

(c) The C–H correlation links the carbon signals with their attached protons and also identifies how many –CH–, –CH$_2$–, –CH$_3$ and quaternary carbons are in the molecule.

(iii) **Analyse the INADEQUATE spectrum** if one is provided, because this can sequentially provide the whole carbon skeleton of the molecule. Choose one signal as a starting point and sequentially work through the INADEQUATE spectrum to determine which carbons are connected to which.

(iv) **Analyse the HMBC spectrum.** This is perhaps the most useful technique to pull together all of the fragments of a molecule because it gives long-range connectivity.

(v) **Analyse the NOESY spectrum** to assign any stereochemistry in the structure.

(vi) **Continually update the list of structural elements** or fragments that have been conclusively identified at each step and start to pull together reasonable possible structures. Be careful not to jump to possible solutions before the evidence is conclusive. Keep assessing and re-assessing all of the options.

(vii) When you have a final solution which you believe is correct, **go back and confirm that all of the spectroscopic data are consistent with the final structure** and that every peak in every spectrum can be properly rationalised in terms of the structure that you have proposed.

L. D. Field
H. L. Li
A. M. Magill
January 2015

1	I—CH₂—CH₂—CH₃ 1-iodopropane C₃H₇I	LABEL COSY HSQC HMBC INADEQUATE	**7**	CH₃CH₂O───────Cl O 3-ethoxypropionyl chloride C₅H₉ClO₂	COSY HSQC HMBC ISOMER (2)

2	CH₃—C—CH₂—CH₃
	      ‖
	      O
	
	 2-butanone
	
	 C₄H₈O
		LABEL
		COSY
		HSQC
		HMBC
		INADEQUATE
			8	O
				 ‖
				 Cl───────O—CH₂CH₃
				
				 ethyl 3-chloropropionate
				
				 C₅H₉ClO₂
					COSY
					HSQC
					HMBC
					ISOMER (2)

3	CH₃—C—CH₂–CH₂–CH₂–CH₃
	     ‖
	     O
	
	 2-hexanone
	
	 C₆H₁₂O
		COSY
		HSQC
		HMBC
			9	CH₃ O
				             ‖
				 CH₃ ─── O—C—CH₃
				
				 isoamyl acetate
				
				 C₇H₁₄O₂
					LABEL
					COSY
					HSQC
					HMBC
					INADEQUATE

4	O
	 ‖
	 CH₃CH₂—C—O—CH₂CH₃
	
	 ethyl propionate
	
	 C₅H₁₀O₂
		SIMULATE
		COSY
		HSQC
		HMBC
			10	O H
				 ‖ |
				 CH₃CH₂—C—C=C—CH₃
				             |
				             H
				
				 trans-4-hexen-3-one
				
				 C₆H₁₀O
					LABEL
					COSY
					HSQC
					HMBC
					NOESY

5	CH₃CH₂O─────OCH₂CH₃
	                  O
	
	 ethyl 3-ethoxypropionate
	
	 C₇H₁₄O₃
		LABEL
		COSY
		HSQC
		HMBC
			11	H
				       |
				 CH₃—C=C────────CH₃
				         | ‖
				        H O
				
				 trans-2-octen-4-one
				
				 C₈H₁₄O
					COSY
					HSQC
					HMBC
					NOESY

6	CH₃ ─────── OH
	    ‖ ‖
	    O O
	
	 4-acetylbutyric acid
	
	 C₆H₁₀O₃
		ASSIGNMENT
		HSQC
		HMBC
			12	H─C=O
				
				 NO₂
				
				 3-nitrobenzaldehyde
				
				 C₇H₅NO₃
					LABEL
					COSY
					HSQC
					HMBC
					NOESY
					INADEQUATE

Organic Structures from 2D NMR Spectra

13	CH₃ ... 3-iodotoluene C_7H_6I	SIMULATE HSQC HMBC	19	CH₃—⬡—C(=O)—O—⬡ phenyl *p*-toluate $C_{14}H_{12}O_2$	COSY HSQC HMBC ISOMER (4)
14	NO₂ ... OH 8-hydroxy-5-nitroquinoline $C_9H_6N_2O_3$	LABEL COSY HSQC HMBC INADEQUATE	20	⬡—⬡—O—C(=O)—CH₃ 4-biphenylyl acetate $C_{14}H_{12}O_2$	COSY HSQC HMBC ISOMER (4)
15	CH₃ ... N Br 2-bromo-3-picoline C_6H_6BrN	HSQC HMBC	21	⬡—O—⬡—C(=O)—CH₃ 4'-phenoxyacetophenone $C_{14}H_{12}O_2$	COSY HSQC HMBC ISOMER (4)
16	H CH₃ ... O—CH₃ *trans*-anethole $C_{10}H_{12}O$	SIMULATE COSY NOESY	22	(CH₃)₃C—⬡—C(=O)—CH₃ 4'-*tert*-butylacetophenone $C_{12}H_{16}O$	HSQC HMBC ISOMER (2)
17	H H CH₃ CH₂—CH₃ *cis*-2-pentene C_5H_{10}	COSY HSQC NOESY	23	CH₃—⬡—C(=O)—C(CH₃)₃ 2,2,4'-trimethyl-propiophenone $C_{12}H_{16}O$	HSQC HMBC ISOMER (2)
18	CH₃—⬡—O—C(=O)—⬡ *p*-tolyl benzoate $C_{14}H_{12}O_2$	COSY HSQC HMBC ISOMER (4)	24	H ... CH₂OH CH₃ *trans*-2-methyl-3-phenyl-2-propen-1-ol $C_{10}H_{12}O$	COSY HSQC HMBC NOESY

25	methyl 4-ethoxybenzoate $C_{10}H_{12}O_3$	COSY HSQC HMBC	31	*trans*-2,*cis*-6-nonadienal $C_9H_{14}O$	COSY HSQC HMBC NOESY
26	methyl 3-(*p*-tolyl)propionate $C_{11}H_{14}O_2$	COSY HSQC HMBC ISOMER (2)	32	allyl glycidyl ether $C_6H_{10}O_2$	COSY HSQC HMBC ISOMER (2)
27	4-(4'-methoxyphenyl)-2-butanone $C_{11}H_{14}O_2$	HSQC INADEQUATE ISOMER (2)	33	3,4-epoxy-4-methyl-2-pentanone $C_6H_{10}O_2$	HSQC HMBC ISOMER (2)
28	ethyl 6-bromohexanoate $C_8H_{15}BrO_2$	ASSIGNMENT COSY HSQC	34	*dl*-methionine $C_5H_{11}NO_2S$	IDENTIFY 1 HSQC HMBC
29	piperonal $C_8H_6O_3$	SIMULATE HSQC HMBC	35	*N*-acetyl-*l*-leucine $C_8H_{15}NO_3$	COSY HSQC HMBC
30	*cis*-3-hexenyl benzoate $C_{13}H_{16}O_2$	COSY HSQC HMBC NOESY	36	isoamyl valerate $C_{10}H_{20}O_2$	TOCSY HSQC

3

Organic Structures from 2D NMR Spectra

37 (E)-4-methyl-4'-nitrostilbene $C_{15}H_{13}NO_2$	ASSIGNMENT COSY HSQC HMBC
43 trans-ferulic Acid $C_{10}H_{10}O_4$	COSY HSQC HMBC
38 2-tert-butyl-6-methylphenol $C_{11}H_{16}O$	HSQC HMBC
44 seC–butyl 3-hydroxycinnamate $C_{13}H_{16}O_3$	COSY HSQC HMBC
39 2-allyl-6-methylphenol $C_{10}H_{12}O$	COSY HSQC HMBC
45 1-benzosuberone $C_{11}H_{12}O$	ASSIGNMENT COSY HSQC HMBC
40 2-hydroxy-4-methoxy-benzaldehyde $C_8H_8O_3$	HSQC HMBC
46 dimethyl (3-bromo-propyl)phosphonate $C_5H_{12}BrO_3P$	COSY HSQC P-H HMBC HETEROATOM
41 2'-hydroxy-5'-methylacetophenone $C_9H_{10}O_2$	HSQC HMBC
47 caffeine $C_8H_{10}N_4O_2$	ASSIGNMENT HSQC HMBC N-H HSQC HETEROATOM
42 3'-fluoro-4'-methoxyacetophenone $C_9H_9FO_2$	HSQC NOESY HETEROATOM
48 benzyloxypropionitrile $C_{10}H_{11}NO$	COSY HSQC HMBC

49	cineole $C_{10}H_{18}O$	HSQC INADEQUATE	55	ethyl acetamido-cyanoacetate $C_7H_{10}N_2O_3$	COSY HSQC HMBC N-H HSQC N-H HMBC
50	thymoquinone $C_{10}H_{12}O_2$	IDENTIFY 1 HSQC HMBC	56	α-Humulene $C_{15}H_{24}$	HSQC INADEQUATE
51	4-bromo-1-indanol C_9H_9BrO	COSY HSQC HMBC	57	3,4-dihydro-2*H*–benzo-pyran-3-carboxylic acid $C_{10}H_{10}O_3$	COSY HSQC HMBC
52	1-bromo-4-methylnaphthalene $C_{11}H_9Br$	ASSIGNMENT COSY HSQC HMBC	58	quinidine $C_{20}H_{24}N_2O_2$	ASSIGNMENT COSY HSQC HMBC NOESY
53	carvacrol $C_{10}H_{14}O$	COSY HSQC HMBC	59	salbutamol $C_{13}H_{21}NO_3$	COSY HSQC HMBC NOESY N-H HMBC HETEROATOM
54	acetoacetanilide $C_{10}H_{11}NO_2$	HSQC HMBC NOESY	60	2-hydroxy-1-naphthaldehyde $C_{11}H_8O_2$	COSY HSQC HMBC

Organic Structures from 2D NMR Spectra

61 6-methyl-4-chromanone $C_{10}H_{10}O_2$	HSQC HMBC INADEQUATE	
62 citronellal $C_{10}H_{18}O$	COSY HSQC HMBC	
63 (+)-*cis*-2-oxabicyclo- [3.3.0]oct-6-en-3-one $C_7H_8O_2$	COSY HSQC HMBC NOESY	

64 melatonin $C_{13}H_{16}N_2O_2$	COSY HSQC HMBC NOESY N-H HSQC N-H HMBC HETEROATOM	
65 carvone $C_{10}H_{14}O$	COSY HSQC HMBC	
66 haloperidol $C_{21}H_{23}ClFNO_2$	ASSIGNMENT COSY HSQC HMBC HETEROATOM	

Question:

The ^1H and $^{13}C\{^1H\}$ NMR spectra of 1-iodopropane (C_3H_7I) recorded in $CDCl_3$ solution at 298 K and 400 MHz are given below.

The ^1H NMR spectrum has signals at δ 0.99 (H_3), 1.84 (H_2) and 3.18 (H_1) ppm.

The $^{13}C\{^1H\}$ NMR spectrum has signals at δ 9.6 (C_1), 15.3 (C_3) and 26.9 (C_2) ppm.

Also given on the following pages are the ^1H–^1H COSY, ^1H–^{13}C me-HSQC, ^1H–^{13}C HMBC and INADEQUATE spectra. For each 2D spectrum, indicate which correlation gives rise to each cross-peak by placing an appropriate label in the box provided (*e.g.* $H_1 \rightarrow H_2$, $H_1 \rightarrow C_1$).

Solution:

1-Iodopropane

$$\overset{1}{I-CH_2}-\overset{2}{CH_2}-\overset{3}{CH_3}$$

1. ^1H–^1H COSY spectra show which pairs of protons are coupled to each other. The COSY spectrum is always symmetrical about a diagonal. In the COSY spectrum, there are two $^3J_{H-H}$ correlations above the diagonal ($H_1 \rightarrow H_2$ and $H_2 \rightarrow H_3$). There are no long-range correlations.

^1H–^1H COSY spectrum of 1-iodopropane ($CDCl_3$, 400 MHz)

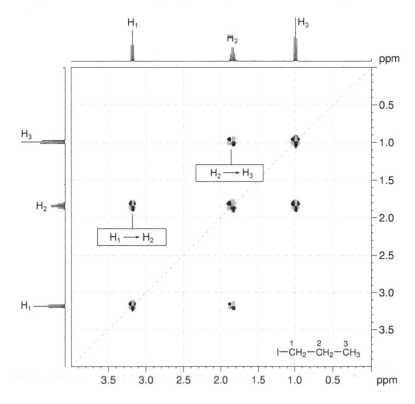

2. The ^{1}H–^{13}C me-HSQC spectrum shows direct (one-bond) correlations between proton and carbon nuclei, so there will be cross-peaks between H_1 and C_1, H_2 and C_2 and also between H_3 and C_3. As the spectrum is multiplicity edited, the cross-peaks corresponding to CH_2 groups are shown in red and are of opposite phase to those for CH_3 groups.

^{1}H–^{13}C me-HSQC spectrum of 1-iodopropane (CDCl₃, 400 MHz)

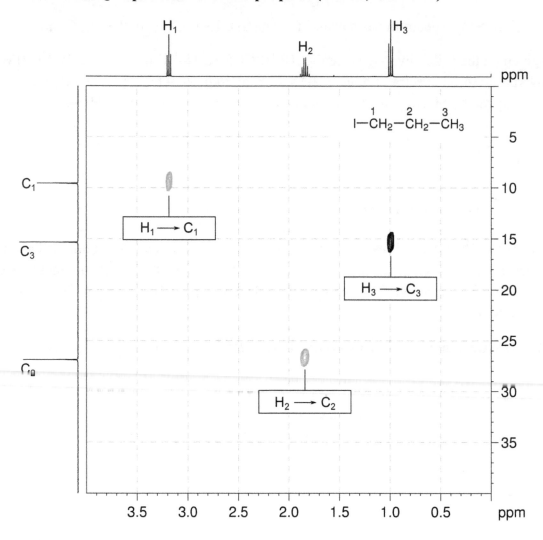

3. In HMBC spectra, remember that, for alkyl systems, both two- and three-bond C–H coupling can give rise to strong cross-peaks.

4. H_1 correlates to C_2 and C_3. H_2 correlates to C_1 and C_3. H_3 correlates to C_1 and C_2.

^1H–^{13}C HMBC spectrum of 1-iodopropane (CDCl$_3$, 400 MHz)

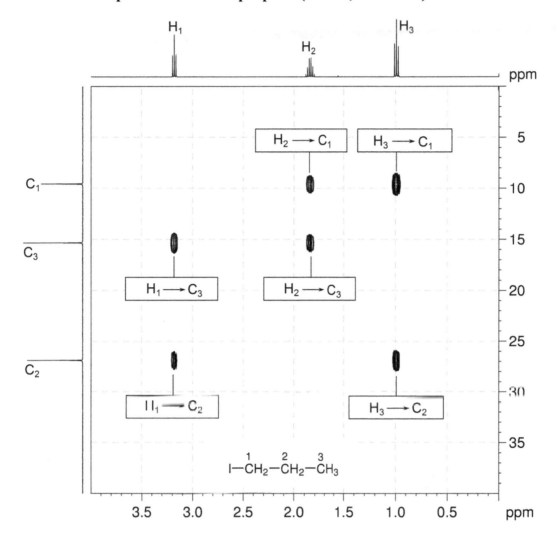

5. The INADEQUATE spectrum shows one-bond ^{13}C–^{13}C connectivity. There are correlations between C_1 and C_2, and C_2 and C_3.

INADEQUATE spectrum of 1-iodopropane (CDCl₃, 150 MHz)

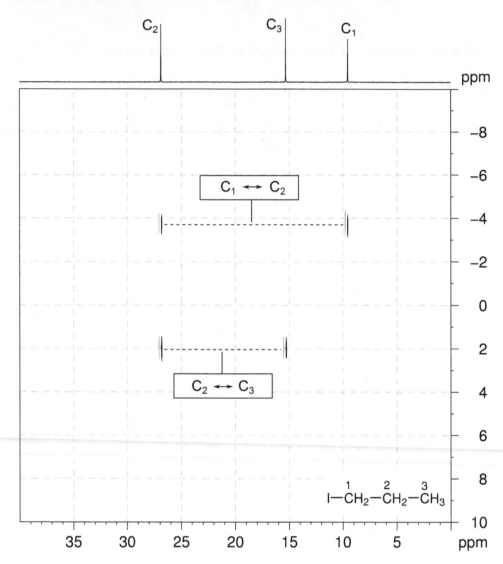

Problem 2

Question:

The ^1H and $^{13}C\{^1H\}$ NMR spectra of 2-butanone (C_4H_8O) recorded in $CDCl_3$ solution at 298 K and 400 MHz are given below.

The ^1H NMR spectrum has signals at δ 1.05 (H_4), 2.14 (H_1) and 2.47 (H_3) ppm.

The $^{13}C\{^1H\}$ NMR spectrum has signals at δ 7.2 (C_4), 28.8 (C_1), 36.2 (C_3) and 208.8 (C_2) ppm.

Also given on the following pages are the ^1H–^1H COSY, ^1H–^{13}C me-HSQC, ^1H–^{13}C HMBC and INADEQUATE spectra. For each 2D spectrum, indicate which correlation gives rise to each cross-peak by placing an appropriate label in the box provided (*e.g.* $H_1 \rightarrow H_2$, $H_1 \rightarrow C_1$).

Solution:

2-Butanone

$$\overset{1}{C}H_3-\overset{2}{\underset{\underset{O}{\|}}{C}}-\overset{3}{C}H_2-\overset{4}{C}H_3$$

1. ^1H–^1H COSY spectra show which pairs of protons are coupled to each other. The COSY spectrum is always symmetrical about a diagonal. In the COSY spectrum, there is only one $^3J_{H-H}$ correlation above the diagonal ($H_3 \rightarrow H_4$). There are no long-range correlations.

^1H–^1H COSY spectrum of 2-butanone ($CDCl_3$, 400 MHz)

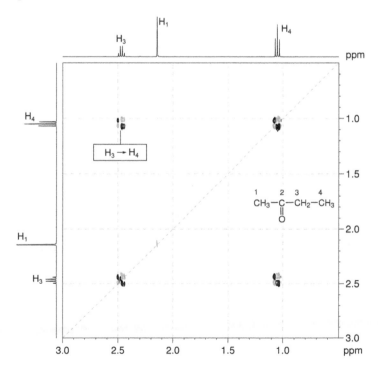

2. The ^{1}H–^{13}C me-HSQC spectrum shows direct (one-bond) correlations between proton and carbon nuclei, so there will be cross-peaks between H_1 and C_1, H_3 and also between C_3 and H_4 and C_4. As the spectrum is multiplicity edited, the cross-peaks corresponding to CH$_2$ groups are shown in red and are of opposite phase to those for CH$_3$ groups.

^{1}H–^{13}C me-HSQC spectrum of 2-butanone (CDCl$_3$, 400 MHz)

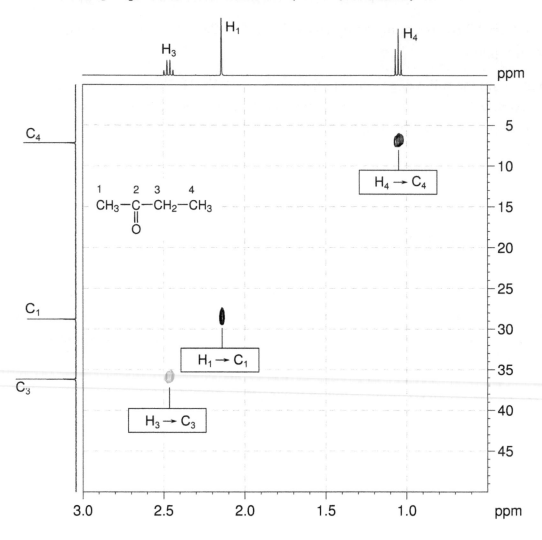

3. In HMBC spectra, remember that, for alkyl systems, both two- and three-bond coupling can give rise to strong cross-peaks. There are no one-bond C–H correlations.

4. H_1 correlates to C_2 and C_3. H_3 correlates to C_1, C_2 and C_4. H_4 correlates to C_2 and C_3.

^{1}H–^{13}C HMBC spectrum of 2-butanone (CDCl$_3$, 400 MHz)

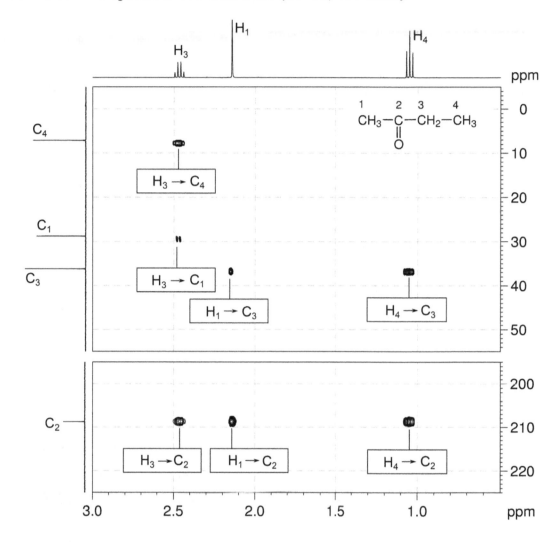

5. The INADEQUATE spectrum shows one-bond ^{13}C–^{13}C connectivity. There are correlations between C_1 and C_2, C_2 and C_3 and C_3 and C_4.

INADEQUATE spectrum of 2-butanone (CDCl₃, 150 MHz)

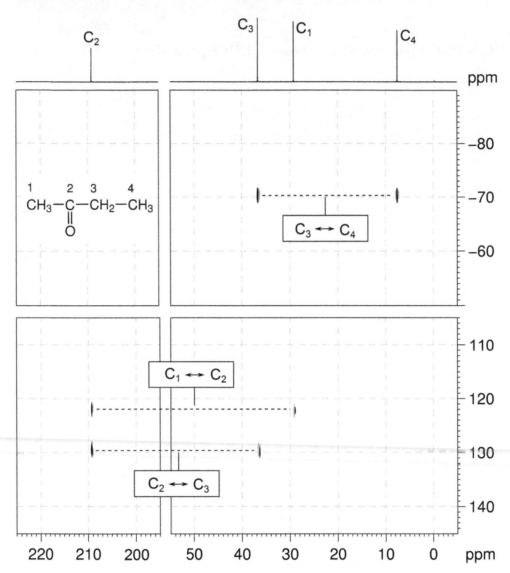

Question:

Identify the following compound.

Molecular Formula: $C_6H_{12}O$

IR: 1718 cm^{-1}

Solution:

2-Hexanone

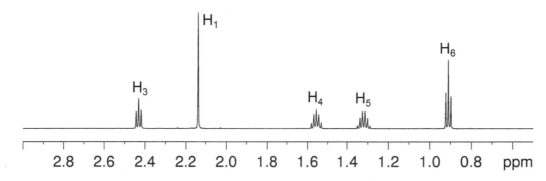

$$\underset{1}{CH_3}-\underset{2}{\underset{\underset{O}{\|}}{C}}-\underset{3}{CH_2}-\underset{4}{CH_2}-\underset{5}{CH_2}-\underset{6}{CH_3}$$

1. The molecular formula is $C_6H_{12}O$. Calculate the degree of unsaturation from the molecular formula: ignore the O atom to give an effective molecular formula of C_6H_{12} (C_nH_m) which gives the degree of unsaturation as $(n - m/2 + 1) = 6 - 6 + 1 = 1$. The compound contains one ring or one functional group containing a double bond.

2. The $^{13}C\{^1H\}$ spectrum establishes that the compound contains a ketone (^{13}C resonance at 209.3 ppm). There can be no other double bonds or rings in the molecule because the C=O accounts for the single degree of unsaturation.

3. 1D NMR spectra establish the presence of three CH_2 groups and two CH_3 groups. The multiplicities of the signals can be verified using the me-HSQC spectrum.

^1H NMR spectrum of 2-hexanone (CDCl$_3$, 600 MHz)

$^{13}C\{H\}$ NMR spectrum of 2-hexanone (CDCl$_3$, 150 MHz)

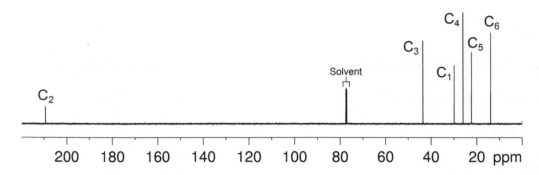

4. The COSY spectrum shows a single spin system – $H_3 \rightarrow H_4$, $H_4 \rightarrow H_5$ and $H_5 \rightarrow H_6$ for a $-CH_2CH_2CH_2CH_3$ fragment.

5. H_1 does not couple to any of the other protons in the molecule and therefore does not show any correlations in the COSY spectrum.

$^1H-^1H$ COSY spectrum of 2-hexanone (CDCl$_3$, 600 MHz)

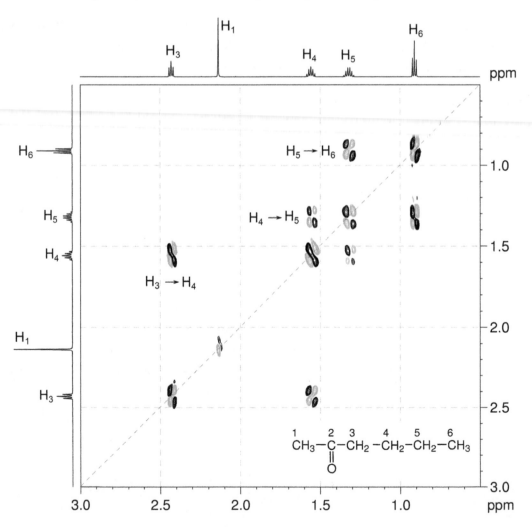

5. The ^{1}H–^{13}C me-HSQC spectrum easily identifies the protonated carbon resonances: C_6 at 13.9, C_5 at 22.4, C_4 at 26.0, C_1 at 29.9 and C_3 at 43.5 ppm.

^{1}H–^{13}C me-HSQC spectrum of 2-hexanone (CDCl₃, 600 MHz)

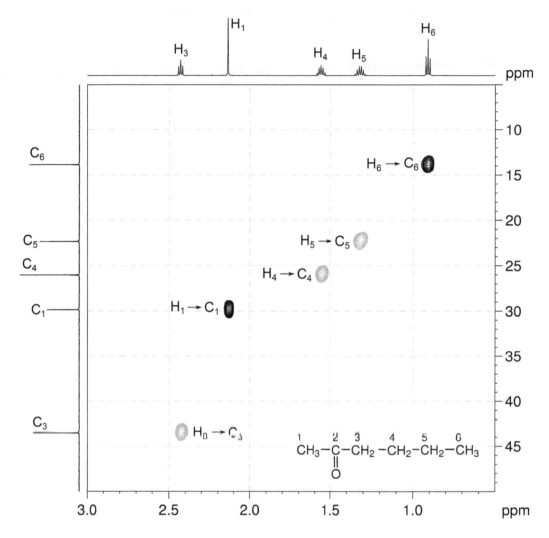

6. The HMBC spectrum confirms the structure with correlations from H_1 and H_3 to C_2 indicating that the ketone group is located between C_1 and C_3. All other correlations are consistent with the structure.

^1H–^{13}C HMBC spectrum of 2-hexanone (CDCl$_3$, 600 MHz)

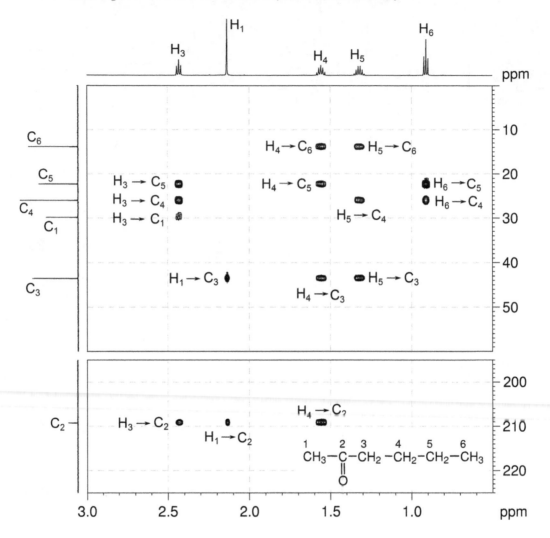

Problem 4

Question:

The ^1H and $^{13}C\{^1H\}$ NMR spectra of ethyl propionate ($C_5H_{10}O_2$) recorded in $CDCl_3$ solution at 298 K and 300 MHz are given below.

The ^1H NMR spectrum has signals at δ 1.14 (H_1), 1.26 (H_5), 2.31 (H_2) and 4.12 (H_4) ppm.

The $^{13}C\{^1H\}$ NMR spectrum has signals at δ 9.2 (C_1), 14.3 (C_5), 27.7 (C_2), 60.3 (C_4) and 174.5 (C_3) ppm.

Use this information to produce schematic diagrams of the COSY, HSQC and HMBC spectra, showing where all of the cross-peaks and diagonal peaks would be.

Solution:

Ethyl propionate

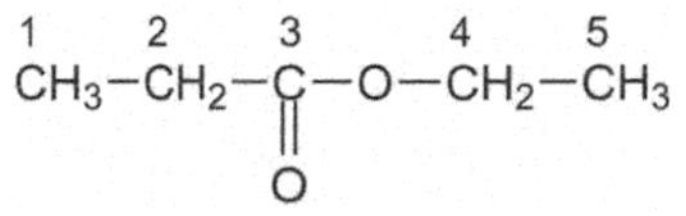

$$\begin{matrix} 1 & 2 & 3 & & 4 & 5 \\ CH_3 - CH_2 - \underset{\underset{O}{\|}}{C} - O - CH_2 - CH_3 \end{matrix}$$

1. The molecule contains two independent spin systems – one for each CH_2CH_3 fragment. Each spin system is made up of two unique spins – one CH_2 and one CH_3.

2. The COSY spectrum has peaks on the diagonal for each unique spin, so the spectrum will contain four diagonal peaks.

3. COSY spectra show cross-peaks (off-diagonal peaks) at positions where a proton whose resonance appears on the horizontal axis is directly coupled to another whose resonance appears on the vertical axis.

4. For ethyl propionate, the CH_2 of each spin system will couple to the CH_3 of the same spin system, so two cross-peaks would be expected – one between H_4 and H_5, and another between H_1 and H_2.

5. Remember that a COSY spectrum is symmetrical about the diagonal, so the two peaks above the diagonal must also be reflected below the diagonal.

Predicted ¹H–¹H COSY spectrum of ethyl propionate (CDCl₃, 300 MHz)

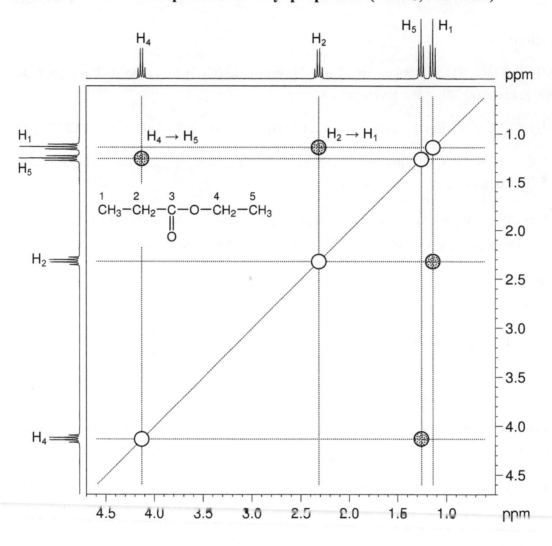

6. The HSQC spectrum contains cross-peaks at positions where a proton whose resonance appears on the horizontal axis is directly bound to a carbon atom whose resonance appears on the vertical axis. There are four cross-peaks in the HSQC spectrum.

Predicted ^{1}H–^{13}C HSQC spectrum of ethyl propionate (CDCl$_3$, 300 MHz)

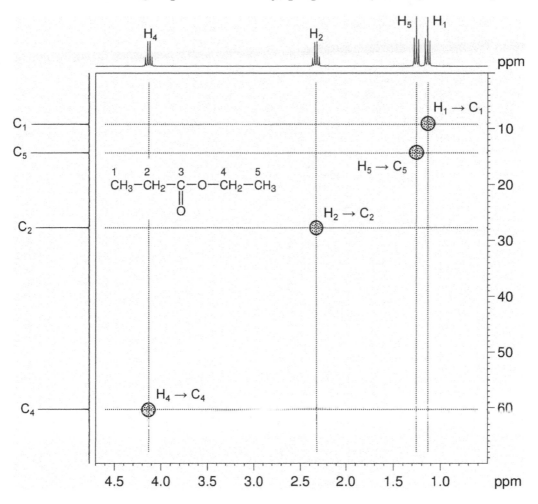

7. The HMBC spectrum contains cross-peaks at positions where a proton whose resonance appears on the horizontal axis is separated by two or three bonds from a carbon atom whose resonance appears on the vertical axis.

Predicted ¹H–¹³C HMBC spectrum of ethyl propionate (CDCl₃, 300 MHz)

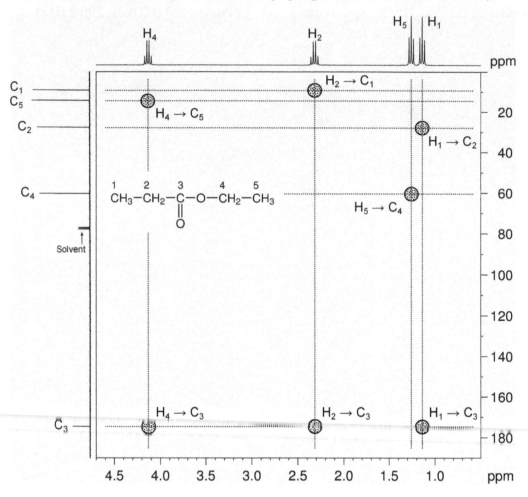

Problem 5

Question:

The ^{1}H and ^{13}C{^{1}H} NMR spectra of ethyl 3-ethoxypropionate ($C_7H_{14}O_3$) recorded in CDCl$_3$ solution at 298 K and 600 MHz are given below.

The ^{1}H NMR spectrum has signals at δ 1.18 (H$_1$), 1.26 (H$_7$), 2.56 (H$_4$), 3.50 (H$_2$), 3.70 (H$_3$) and 4.15 (H$_6$) ppm.

The ^{13}C{^{1}H} NMR spectrum has signals at δ 14.2 (C$_7$), 15.1 (C$_1$), 35.3 (C$_4$), 60.4 (C$_6$), 65.9 (C$_3$), 66.4 (C$_2$) and 171.7 (C$_5$) ppm.

Also given on the following pages are the ^{1}H–^{1}H COSY, ^{1}H–^{13}C me-HSQC and ^{1}H–^{13}C HMBC spectra. For each 2D spectrum, indicate which correlation gives rise to each cross-peak by placing an appropriate label in the box provided (*e.g.* H$_1$ → H$_2$, H$_1$ → C$_1$).

Solution:

Ethyl 3-ethoxypropionate

$$\overset{1}{C}H_3-\overset{2}{C}H_2-O-\overset{3}{C}H_2-\overset{4}{C}H_2-\overset{5}{\underset{\underset{O}{\|}}{C}}-O-\overset{6}{C}H_2-\overset{7}{C}H_3$$

1. ^{1}H–^{1}H COSY spectra show which pairs of protons are coupled to each other. The COSY spectrum is always symmetrical about a diagonal. In the COSY spectrum, there are three $^3J_{H–H}$ correlations above the diagonal (H$_2$ → H$_1$, H$_3$ → H$_4$ and H$_6$ → H$_7$). There are no long-range correlations.

^{1}H–^{1}H COSY spectrum of ethyl 3-ethoxypropionate (CDCl$_3$, 600 MHz)

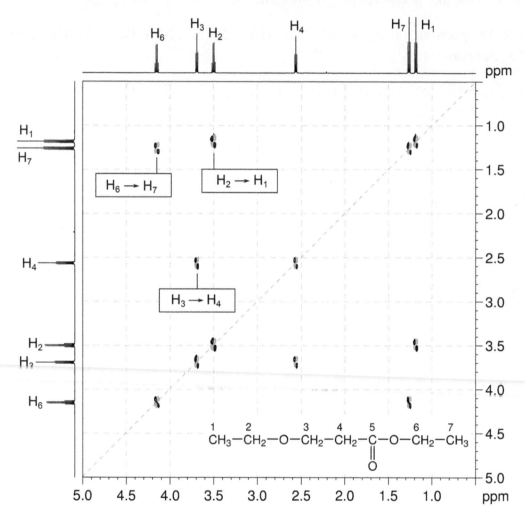

2. The ^1H–^{13}C me-HSQC spectrum shows direct (one-bond) correlations between proton and carbon nuclei, so there will be cross-peaks between H_1 and C_1, H_2 and C_2, H_3 and C_3, H_4 and C_4, H_6 and C_6 and H_7 and C_7. As the spectrum is multiplicity edited, the cross-peaks corresponding to CH_2 groups are shown in red and are of opposite phase to those for CH_3 groups.

^1H–^{13}C me-HSQC spectrum of ethyl 3-ethoxypropionate (CDCl$_3$, 600 MHz)

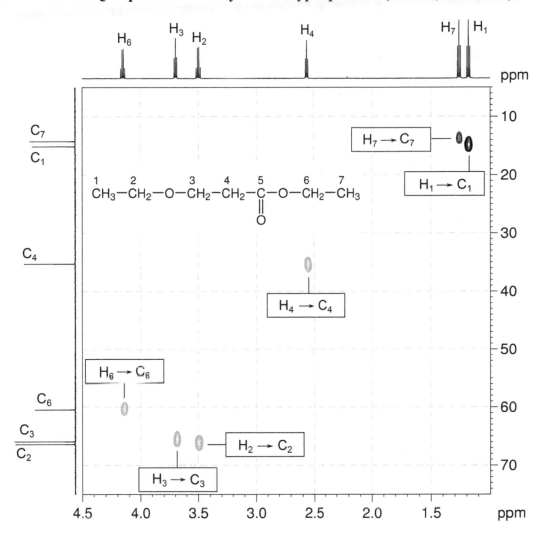

3. In HMBC spectra, remember that, for alkyl systems, both two- and three-bond couplings can give rise to strong cross-peaks.

$^1H–^{13}C$ HMBC spectrum of ethyl 3-ethoxypropionate (CDCl$_3$, 600 MHz)

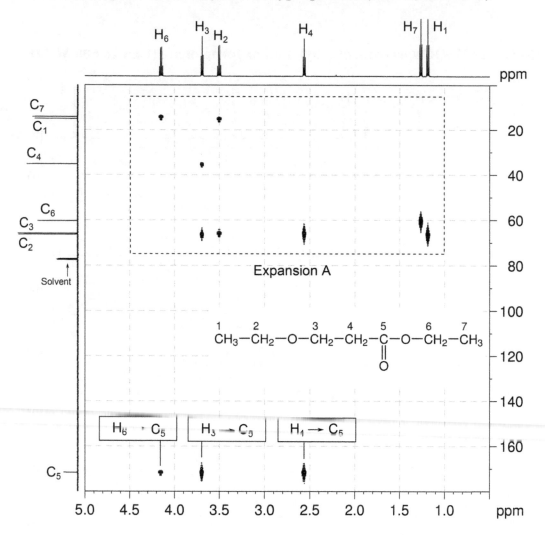

4. H_1 correlates to C_2 only. H_2 correlates to C_1 and C_3. H_3 correlates to C_2, C_4 and C_5. H_4 correlates to C_3 and C_5. H_6 correlates to C_5 and C_7. H_7 correlates to C_6 only.

^{1}H–^{13}C HMBC spectrum of ethyl 3-ethoxypropionate – expansion A

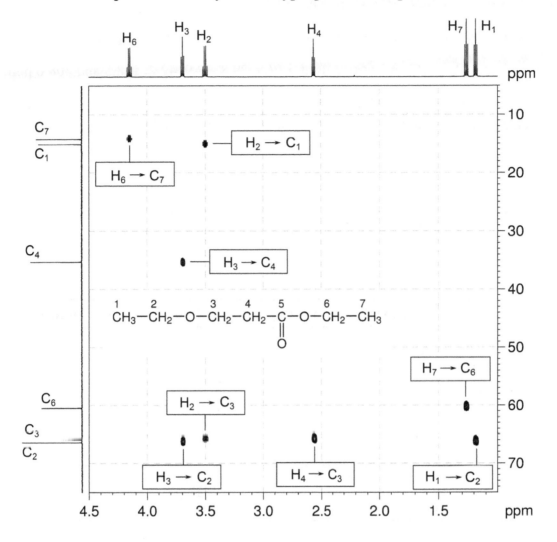

Problem 6

Question:

The ^1H and $^{13}C\{^1H\}$ NMR spectra of 4-acetylbutyric acid ($C_6H_{10}O_3$) recorded in $CDCl_3$ solution at 298 K and 600 MHz are given below.

The ^1H NMR spectrum has signals at δ 1.81, 2.08, 2.31, 2.47 and 10.5 ppm.

The $^{13}C\{^1H\}$ NMR spectrum has signals at δ 18.5, 29.8, 32.9, 42.2, 178.8 and 208.6 ppm.

The 2D me-^1H–^{13}C HSQC and ^1H–^{13}C HMBC spectra are given on the following pages. Use these spectra to assign the ^1H and $^{13}C\{^1H\}$ resonances for this compound.

Solution:

4-Acetylbutyric Acid

$$\underset{O}{\overset{6}{CH_3}}-\underset{\underset{O}{\|}}{\overset{5}{C}}-\overset{4}{CH_2}-\overset{3}{CH_2}-\overset{2}{CH_2}-\underset{\underset{O}{\|}}{\overset{1}{C}}-\overset{}{OH}$$

Proton	Chemical Shift (ppm)	Carbon	Chemical Shift (ppm)
		C_1	178.8
H_2	2.31	C_2	32.9
H_3	1.81	C_3	18.5
H_4	2.47	C_4	42.2
		C_5	208.6
H_6	2.08	C_6	29.8
OH	10.5		

1. The methyl group (H_6, singlet at 2.08 ppm) and the –OH proton (10.5 ppm, exchangeable) may be easily identified from the ^1H NMR spectrum. The triplet resonances (at 2.31 and 2.47 ppm) correspond to H_2 and H_4 but cannot be assigned by inspection. The more complex resonance at 1.81 ppm must correspond to H_3 since the multiplet structure shows it has more than two neighbouring protons.

^{1}H NMR spectrum of 4-acetylbutyric acid (CDCl$_3$, 600 MHz)

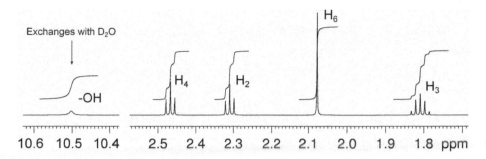

2. The signals corresponding to the ketone (C$_5$, 208.6 ppm) and the carboxylic acid (C$_1$, 178.8 ppm) in the ^{13}C{^{1}H} NMR spectrum can be assigned by inspection.

^{13}C{^{1}H} NMR spectrum of 4-acetylbutyric acid (CDCl$_3$, 150 MHz)

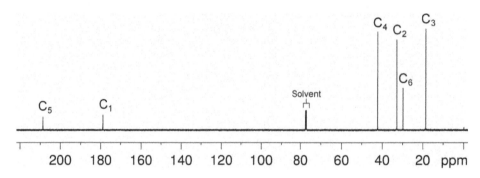

3. The ^1H–^{13}C me-HSQC spectrum easily identifies the protonated carbon resonances: C_3 at 18.5, and C_6 at 29.8 ppm. The proton resonance at 2.31 ppm correlates to the ^{13}C resonance at 32.9 ppm, and the ^1H resonance at 2.47 ppm correlates to the ^{13}C resonance at 42.2 ppm.

^1H–^{13}C me-HSQC spectrum of 4-acetylbutyric acid (CDCl$_3$, 600 MHz)

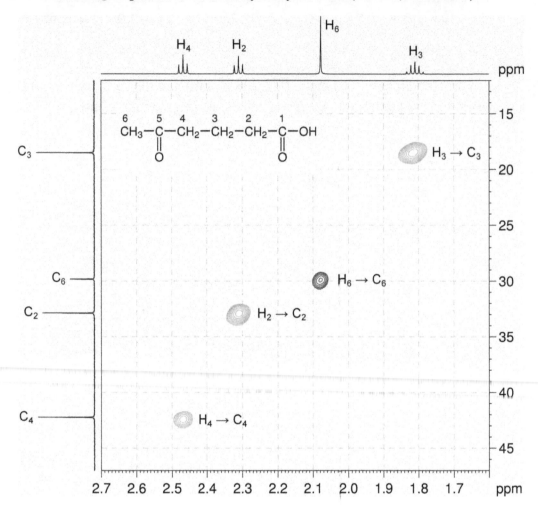

4. Remember that, in HMBC spectra, two- and three-bond correlations are generally strongest in aliphatic systems.

5. In the $^1H-^{13}C$ HMBC spectrum, H_6 correlates to the ketone carbon (C_5, a two-bond correlation), as well as the resonance at 42.2 ppm. This correlation must be the three-bond correlation to C_4, and so we can assign the resonances at 2.47 and 42.2 ppm to H_4 and C_4, respectively. The newly assigned H_4 shows a strong correlation to C_5, confirming its assignment.

6. H_2 must therefore be the resonance at 2.31 ppm, and C_2 the resonance at 32.9 ppm.

7. In the HMBC spectrum, H_2 correlates strongly to C_1, confirming its assignment.

$^1H-^{13}C$ HMBC spectrum of 4-acetylbutyric acid (CDCl$_3$, 600 MHz)

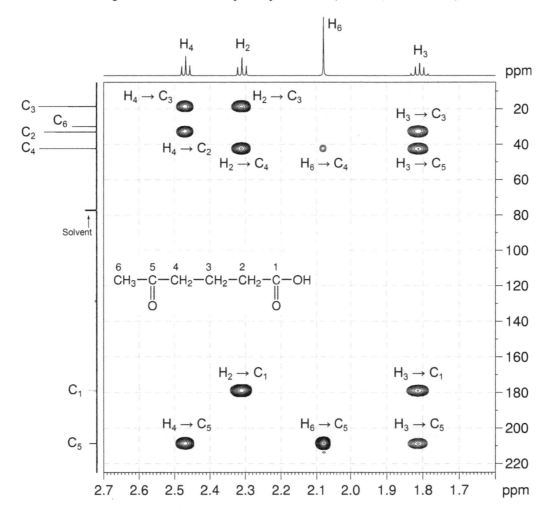

Problem 7

Question:

Identify the following compound.

Molecular Formula: $C_5H_9ClO_2$

Solution:

3-Ethoxypropionyl chloride

$$\overset{5}{CH_3}-\overset{4}{CH_2}-O-\overset{3}{CH_2}-\overset{2}{CH_2}-\overset{1}{C}-Cl$$
$$\quad\quad\quad\quad\quad\quad\quad\quad\quad\quad\;\overset{\|}{O}$$

1. The molecular formula is $C_5H_9ClO_2$. Calculate the degree of unsaturation from the molecular formula: replace the Cl with H and ignore the O atoms to give an effective molecular formula of C_5H_{10} (C_nH_m) which gives the degree of unsaturation as $(n - m/2 + 1) = 5 - 5 + 1 = 1$. The compound contains one ring or one functional group containing a double bond.

2. The $^{13}C\{^1H\}$ spectrum establishes that the compound contains a carbonyl group (^{13}C resonance at 171.9 ppm). This accounts for all of the degrees of unsaturation, so the compound contains no additional rings or multiple bonds.

^1H NMR spectrum of 3-ethoxypropionyl chloride (CDCl₃, 500 MHz)

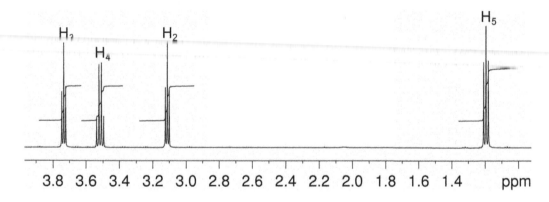

$^{13}C\{^1H\}$ NMR spectrum of 3-ethoxypropionyl chloride (CDCl₃, 125 MHz)

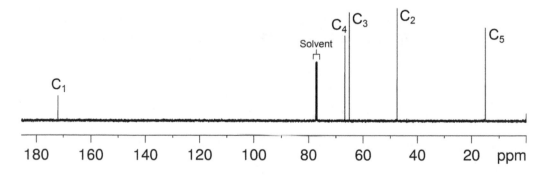

3. The COSY spectrum shows two independent spin systems – H_4 to H_5 for a $-CH_2CH_3$ fragment and H_2 to H_3 for a $-CH_2CH_2-$ fragment.

$^1H-^1H$ COSY spectrum of 3-ethoxypropionyl chloride (CDCl₃, 500 MHz)

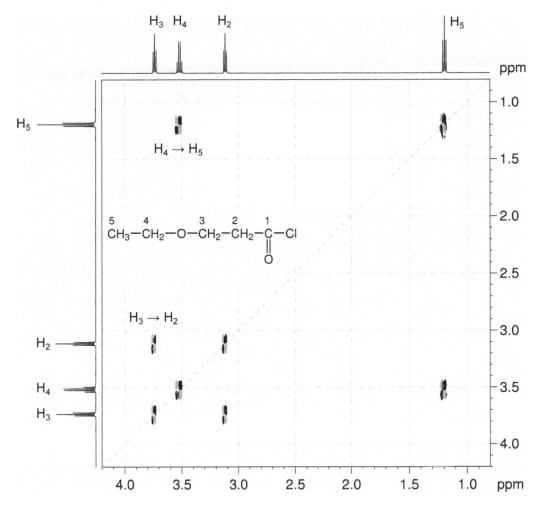

4. In the me-HSQC spectrum, the downfield ^1H and ^{13}C chemical shifts of H_3/C_3 and H_4/C_4 indicate that C_3 and C_4 are bound to either oxygen or chlorine.

^1H–^{13}C me-HSQC spectrum of 3-ethoxypropionyl chloride (CDCl$_3$, 500 MHz)

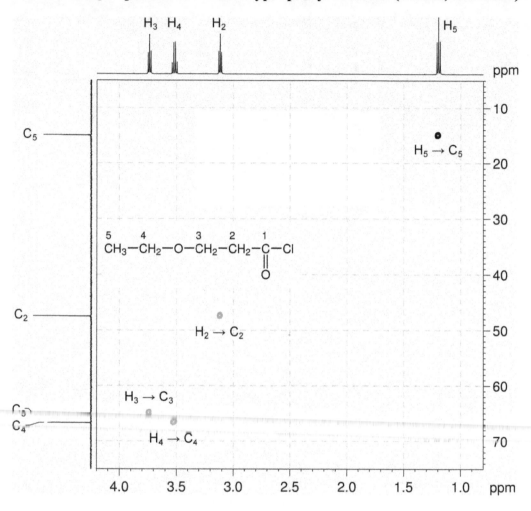

5. There are two possible isomers:

$$Cl-CH_2-CH_2-\underset{\underset{O}{\|}}{C}-O-CH_2-CH_3 \qquad CH_3-CH_2-O-CH_2-CH_2-\underset{\underset{O}{\|}}{C}-Cl$$

A   B

6. In the HMBC spectrum, the correlations from H_2 and H_3 to C_1 indicate that the $-CH_2CH_2-$ fragment is bound to the carbonyl group. The $H_3 \rightarrow C_4$ and $H_4 \rightarrow C_3$ correlations indicate an interaction between the two spin systems. These correlations would be absent in Isomer A thus **Isomer B** is the correct answer.

$^1H-^{13}C$ HMBC spectrum of 3-ethoxypropionyl chloride (CDCl$_3$, 500 MHz)

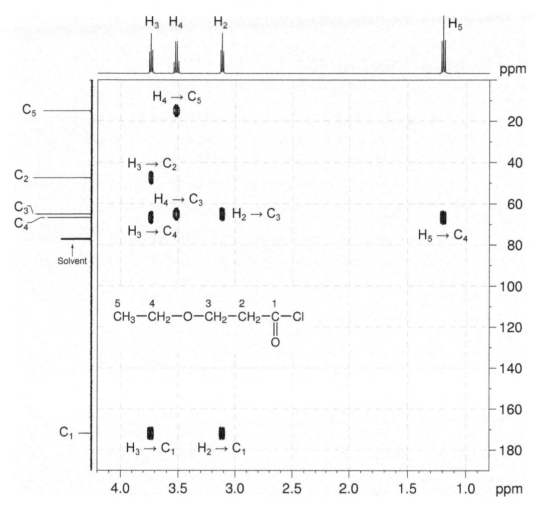

Problem 8

Question:

Identify the following compound.

Molecular Formula: $C_5H_9ClO_2$

Solution:

Ethyl 3-chloropropionate

$$\overset{3}{Cl}-\overset{}{CH_2}-\overset{2}{CH_2}-\overset{1}{\underset{\underset{O}{\|}}{C}}-\overset{4}{O}-\overset{}{CH_2}-\overset{5}{CH_3}$$

1. The molecular formula is $C_5H_9ClO_2$. Calculate the degree of unsaturation from the molecular formula: replace the Cl with H and ignore the O atoms to give an effective molecular formula of C_5H_{10} (C_nH_m) which gives the degree of unsaturation as $(n - m/2 + 1) = 5 - 5 + 1 = 1$. The compound contains one ring or one functional group containing a double bond.

2. The $^{13}C\{^1H\}$ spectrum establishes the presence of a carbonyl group (^{13}C resonance at 170.3 ppm). This accounts for all of the degrees of unsaturation, so the compound contains no additional rings or multiple bonds.

^1H NMR spectrum of ethyl 3-chloropropionate (CDCl₃, 500 MHz)

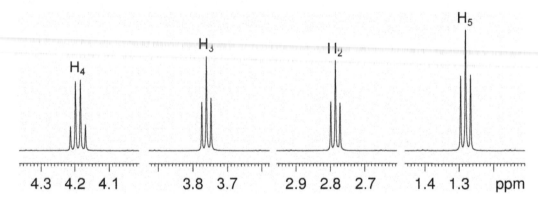

$^{13}C\{^1H\}$ NMR spectrum of ethyl 3-chloropropionate (CDCl₃, 125 MHz)

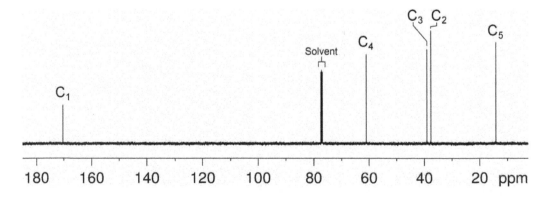

3. The COSY spectrum shows two independent spin systems – H$_4$ to H$_5$ for a –CH$_2$CH$_3$ fragment and H$_2$ to H$_3$ for a –CH$_2$CH$_2$– fragment.

^{1}H–^{1}H COSY spectrum of ethyl 3-chloropropionate (CDCl$_3$, 500 MHz)

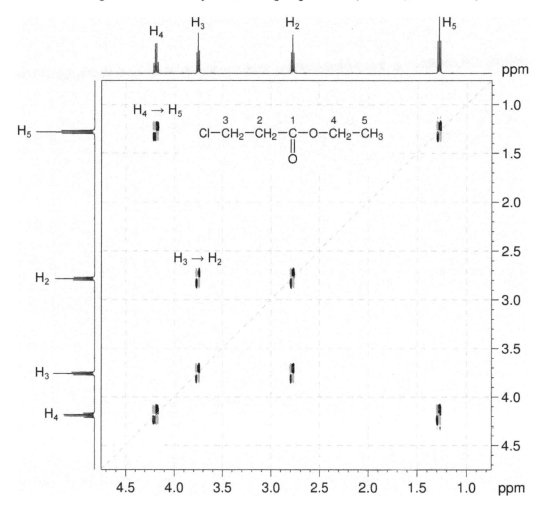

4. In the me-HSQC spectrum, the ^{1}H and ^{13}C chemical shifts of H$_4$ / C$_4$ indicate that C$_4$ is bound to an oxygen atom.

^{1}H–^{13}C me-HSQC spectrum of ethyl 3-chloropropionate (CDCl$_3$, 500 MHz)

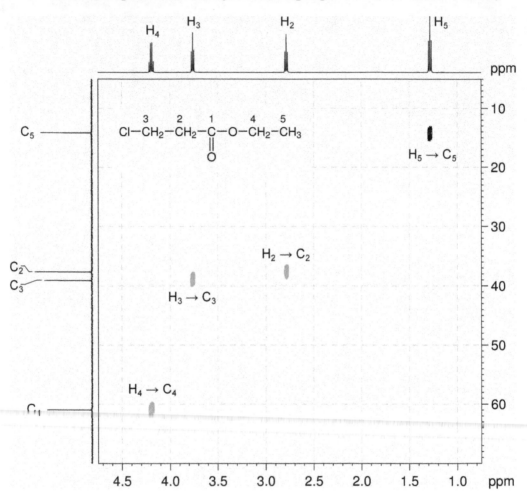

5. The HMBC spectrum shows correlations from H$_4$ to C$_1$ indicating that the carbonyl group must be part of an ester. Correlations from H$_2$ and H$_3$ to C$_1$ indicate that the –CH$_2$CH$_2$– fragment is bound to the carbonyl group leaving the Cl atom to be bound to the other end of the –CH$_2$CH$_2$– fragment.

6. The ^{1}H and ^{13}C chemical shifts of H$_3$ / C$_3$ are consistent with those for a CH$_2$–Cl group.

^{1}H–^{13}C HMBC spectrum of ethyl 3-chloropropionate (CDCl$_3$, 500 MHz)

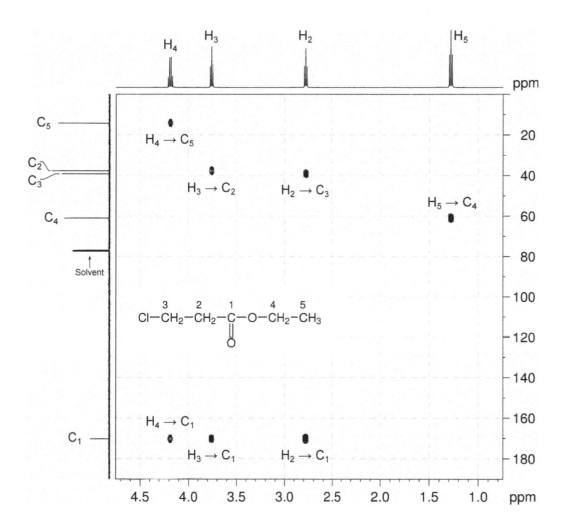

Problem 9

Question:

The ^{1}H and ^{13}C{^{1}H} NMR spectra of isoamyl acetate ($C_7H_{14}O_2$) recorded in CDCl$_3$ solution at 298 K and 600 MHz are given below.

The ^{1}H NMR spectrum has signals at δ 0.92 (H$_6$), 1.52 (H$_4$), 1.69 (H$_5$), 2.04 (H$_1$) and 4.09 (H$_3$) ppm.

The ^{13}C{^{1}H} NMR spectrum has signals at δ 21.0 (C$_1$), 22.5 (C$_6$), 25.1 (C$_5$), 37.4 (C$_4$), 63.1 (C$_3$) and 171.2 (C$_2$) ppm.

Also given on the following pages are the ^{1}H–^{1}H COSY, ^{1}H–^{13}C me-HSQC, ^{1}H–^{13}C HMBC and INADEQUATE spectra. For each 2D spectrum, indicate which correlation gives rise to each cross-peak by placing an appropriate label in the box provided.

Solution:

Isoamyl acetate

$$\underset{\underset{O}{\|}}{\overset{1}{CH_3}-\overset{2}{C}}-O-\overset{3}{CH_2}-\overset{4}{CH_2}-\overset{5}{CH}\overset{\overset{6}{CH_3}}{\underset{\underset{6}{CH_3}}{}}$$

1. ^1H–^1H COSY spectra show which pairs of protons are coupled to each other. The COSY spectrum is always symmetrical about a diagonal. In the COSY spectrum, there are three $^3J_{H-H}$ correlations above the diagonal (H$_3$ → H$_4$; H$_5$ → H$_4$; and H$_5$ → H$_6$). There are no long-range correlations.

^1H–^1H COSY spectrum of isoamyl acetate (CDCl$_3$, 600 MHz)

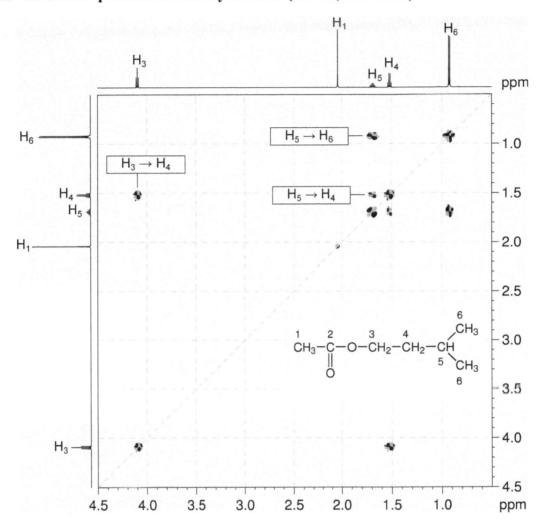

2. The ^{1}H–^{13}C me-HSQC spectrum shows direct (one-bond) correlations between proton and carbon nuclei, so there will be cross-peaks between H_1 and C_1, H_3 and C_3, H_4 and C_4, H_5 and C_5 and also between H_6 and C_6. As the spectrum is multiplicity edited, the cross-peaks corresponding to CH_2 groups are shown in red and are of opposite phase to those for CH and CH_3 groups.

^{1}H–^{13}C me-HSQC spectrum of isoamyl acetate (CDCl$_3$, 600 MHz)

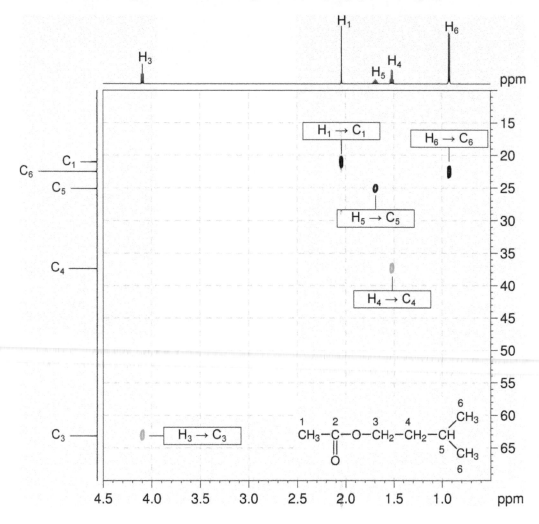

3. In HMBC spectra, remember that, for alkyl systems, both two- and three-bond couplings can give rise to strong cross-peaks.

^{1}H–^{13}C HMBC spectrum of isoamyl acetate (CDCl$_3$, 600 MHz)

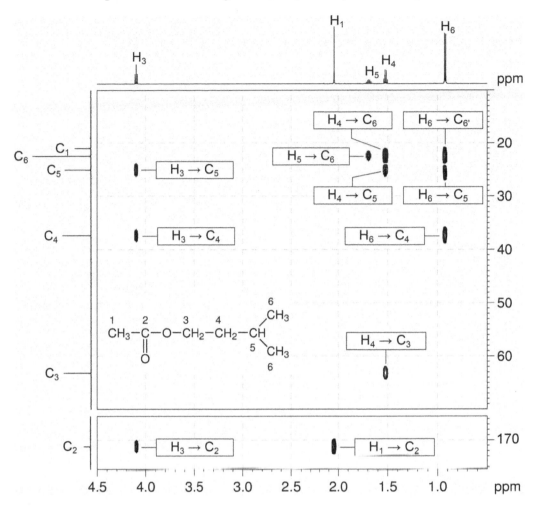

4. H_1 correlates to C_2 only. H_3 correlates to C_2, C_4 and C_5. H_4 correlates to C_3, C_5 and C_6. H_5 correlates to C_6 only. The expected correlations between H_5 and C_4 and H_5 and C_3 are absent from the spectrum.

5. H_6 correlates to C_5. The expected correlation between H_6 and C_4 is absent from the spectrum. There is also a correlation between H_6 and $C_{6'}$. While this appears to be a one-bond correlation, in *gem*-dimethyl groups, the apparent one-bond correlation arises from the $^3J_{C-H}$ interaction of the protons of one of the methyl groups with the chemically equivalent carbon which is three bonds away.

6. The INADEQUATE spectrum shows one-bond $^{13}C-^{13}C$ connectivity. There are correlations between C_1 and C_2, C_3 and C_4, C_4 and C_5 and also between C_5 and C_6. There are no correlations between C_2 and C_3 because of the presence of the oxygen bridge.

INADEQUATE spectrum of isoamyl acetate (CDCl$_3$, 150 MHz)

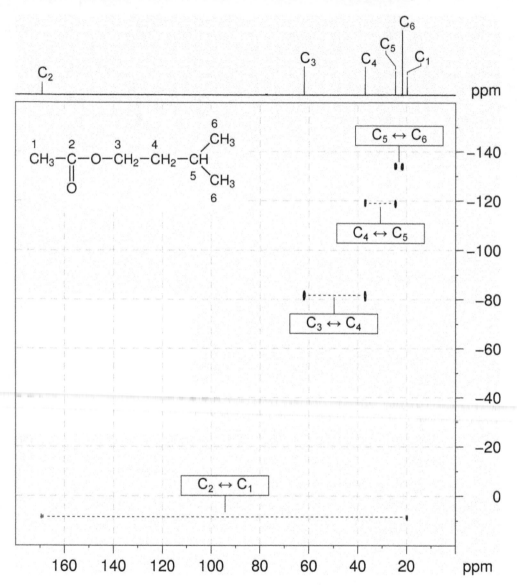

Problem 10

Question:

The ^{1}H and ^{13}C$\{^1$H$\}$ NMR spectra of *trans*-4-hexen-3-one ($C_6H_{10}O$) recorded in DMSO-d_6 solution at 298 K and 400 MHz are given below.

The ^{1}H NMR spectrum has signals at δ 0.96 (H$_1$), 1.86 (H$_6$), 2.56 (H$_2$), 6.11 (H$_4$) and 6.85 (H$_5$) ppm.

The ^{13}C$\{^1$H$\}$ NMR spectrum has signals at δ 8.4 (C$_1$), 18.4 (C$_6$), 32.6 (C$_2$), 131.9 (C$_4$), 142.8 (C$_5$) and 200.4 (C$_3$) ppm.

Also given on the following pages are the ^{1}H–^{1}H COSY, ^{1}H–^{13}C me-HSQC, ^{1}H–^{13}C HMBC and ^{1}H–^{1}H NOESY spectra. For each 2D spectrum, indicate which correlation gives rise to each cross-peak by placing an appropriate label in the box provided.

Solution:

trans-4-Hexen-3-one

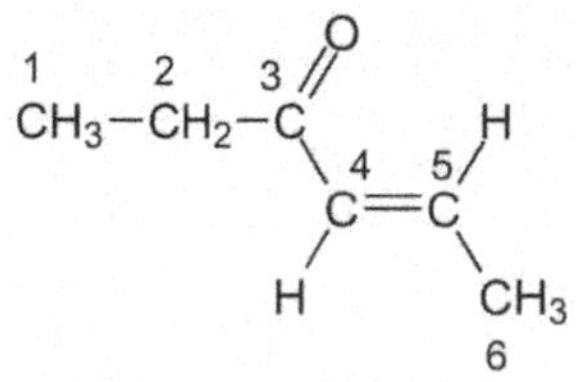

1. $^1H-^1H$ COSY spectra show which pairs of protons are coupled to each other. The COSY spectrum is always symmetrical about a diagonal. In the COSY spectrum, there are four $^3J_{H-H}$ correlations above the diagonal ($H_2 \rightarrow H_1$; $H_4 \rightarrow H_2$; $H_5 \rightarrow H_4$; and $H_5 \rightarrow H_6$).

$^1H-^1H$ COSY spectrum of *trans*-4-hexen-3-one (DMSO-d_6, 400 MHz)

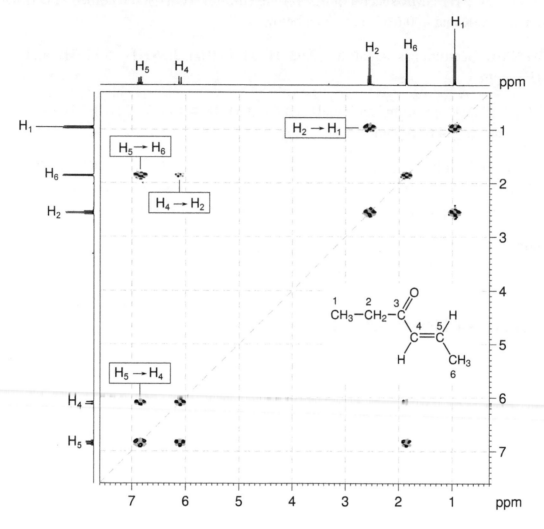

2. The ^1H–^{13}C me-HSQC spectrum shows direct (one-bond) correlations between proton and carbon nuclei, so there will be cross-peaks between H_1 and C_1, H_2 and C_2, H_4 and C_4, H_5 and C_5 and also between H_6 and C_6. As the spectrum is multiplicity edited, the cross-peaks corresponding to CH_2 groups are shown in red and are of opposite phase to those for CH and CH_3 groups.

^1H–^{13}C me-HSQC spectrum of *trans*-4-hexen-3-one (DMSO-d_6, 400 MHz)

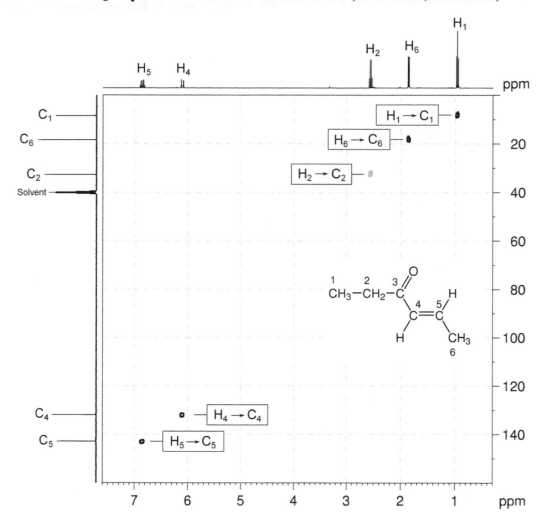

3. In HMBC spectra, remember that, for alkyl systems, both two- and three-bond couplings can give rise to strong cross-peaks.

^{1}H–^{13}C HMBC spectrum of *trans*-4-hexen-3-one (DMSO-d_6, 400 MHz)

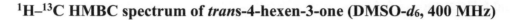

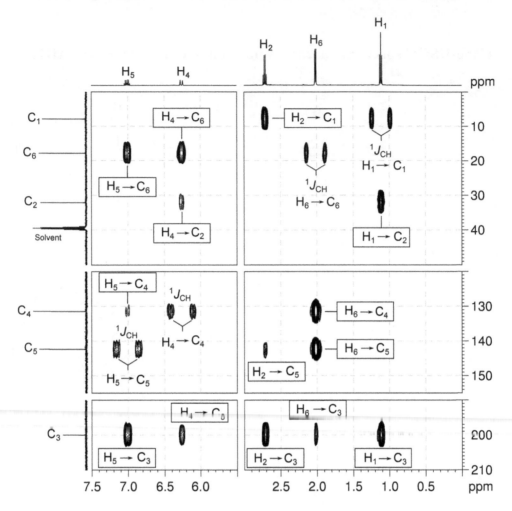

4. H_1 correlates to C_2 and C_3. H_2 correlates to C_1, C_3 and C_5. H_4 correlates to C_2, C_3 and C_6. H_5 correlates to C_3, C_4 and C_6. H_6 correlates to C_3, C_4 and C_5. Note that the $H_2 \rightarrow C_5$ and $H_6 \rightarrow C_3$ correlations are long-range four-bond couplings. The possible $H_2 \rightarrow C_4$, $H_4 \rightarrow C_5$ and $H_5 \rightarrow C_4$ correlations are absent from the spectrum.

5. Note that the one-bond couplings between H_1 and C_1, H_4 and C_4, H_5 and C_5, and H_6 and C_6 are visible in the HMBC spectrum as large doublets.

6. ^{1}H–^{1}H NOESY spectra show the pairs of protons which are close together in space. The NOESY spectrum is always symmetrical about the diagonal. In the NOESY spectrum, there are six correlations above the diagonal (H$_2$ → H$_1$, H$_4$ → H$_2$, H$_5$ → H$_2$, H$_5$ → H$_4$, H$_4$ → H$_6$ and H$_5$ → H$_6$).

7. The H$_4$ → H$_2$ / H$_6$ and H$_5$ → H$_2$ / H$_6$ correlations clearly indicate the *trans* geometry of the alkene functional group in the molecule. The H$_4$ → H$_6$ and H$_5$ → H$_2$ correlations would be absent if the alkene group was of *cis* geometry.

^{1}H–^{1}H NOESY spectrum of *trans*-4-hexen-3-one (DMSO-d_6, 400 MHz)

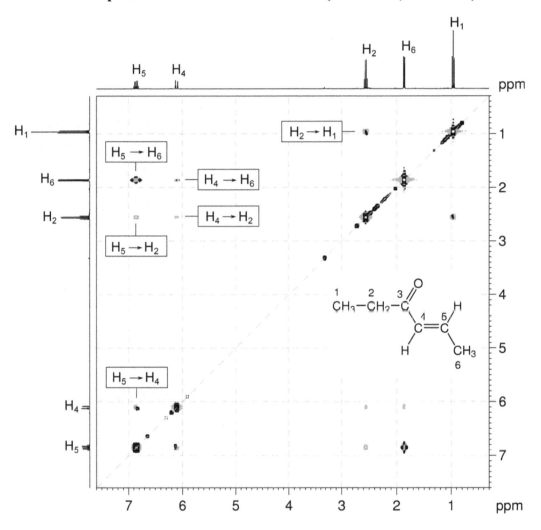

Problem 11

Question:

Identify the following compound.

Molecular Formula: $C_8H_{14}O$

IR: 1698, 1638 cm^{-1}

Solution:

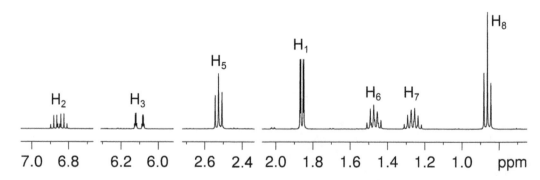

trans-**2-Octen-4-one**

1. The molecular formula is $C_8H_{14}O$. Calculate the degree of unsaturation from the molecular formula: ignore the O atom to give an effective molecular formula of C_8H_{14} (C_nH_m) which gives the degree of unsaturation as $(n - m/2 + 1) = 8 - 7 + 1 = 2$. The compound contains a combined total of two rings and/or π bonds.

2. 1D NMR data establish the presence of an alkene (^1H resonances at 6.85 and 6.10 ppm; ^{13}C resonances at 142.5 and 131.7 ppm) and a ketone functional group (^{13}C resonance at 199.6 ppm). This accounts for all of the degrees of unsaturation, so the compound contains no additional rings or multiple bonds.

3. The ^1H NMR and me-HSQC spectra further establish the presence of two –CH$_3$ and three –CH$_2$ groups.

^1H NMR spectrum of *trans*-2-octen-4-one (DMSO-d_6, 400 MHz)

$^{13}C\{^1H\}$ NMR spectrum of *trans*-2-octen-4-one (DMSO-d_6, 100 MHz)

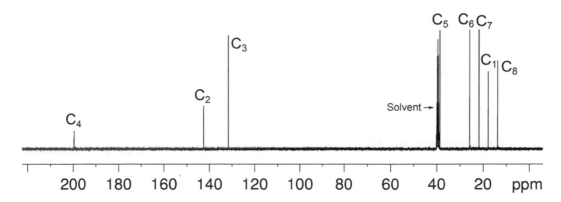

4. The COSY spectrum shows two independent spin systems – H_8 to H_7 to H_6 to H_5 for a $CH_3CH_2CH_2CH_2-$ fragment and H_1 to H_2 to H_3 for a $CH_3CH=CH-$ fragment.

$^1H-^1H$ COSY spectrum of *trans*-2-octen-4-one (DMSO-d_6, 400 MHz)

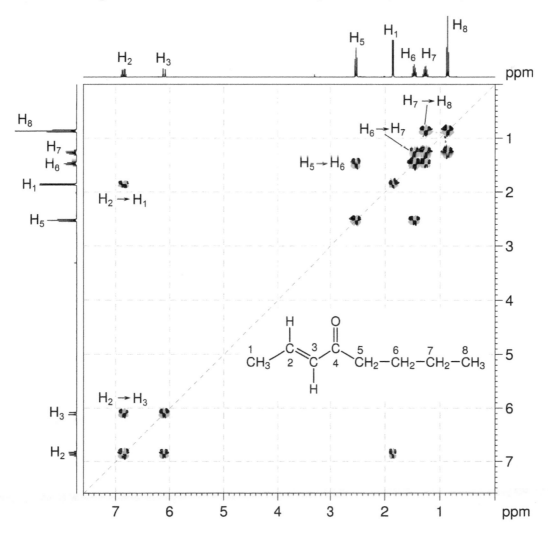

5. The $^{1}H-^{13}C$ me-HSQC spectrum easily identifies the protonated carbon resonances (C_1, C_2, C_3, C_5, C_6, C_7 and C_8).

$^{1}H-^{13}C$ me-HSQC spectrum of *trans*-2-octen-4-one (DMSO-d_6, 400 MHz)

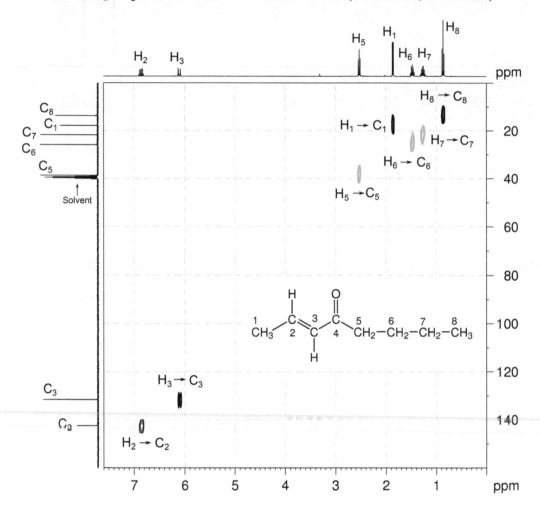

6. The HMBC spectrum shows strong correlations from H_2, H_5 and H_6 to C_4 indicating that the ketone functional group must be located between the $CH_3CH=CH-$ and $CH_3CH_2CH_2CH_2-$ fragments to afford $CH_3CH=CHC(O)CH_2CH_2CH_2CH_3$.

7. The expected $H_2 \rightarrow C_3$, $H_3 \rightarrow C_2$, $H_3 \rightarrow C_5$ and $H_5 \rightarrow C_3$ correlations are absent from the HMBC spectrum. Note that the $H_1 \rightarrow C_4$ correlation is a long-range four-bond coupling. All other correlations are consistent with the structure.

^1H–^{13}C HMBC spectrum of *trans*-2-octen-4-one (DMSO-d_6, 400 MHz)

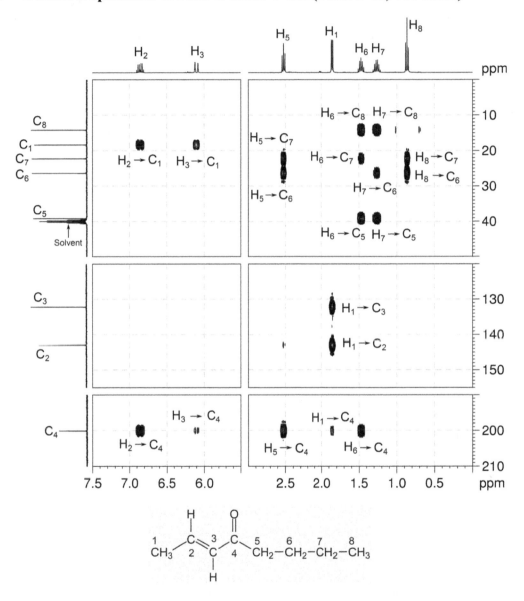

8. ^{1}H–^{1}H NOESY spectra show pairs of protons which are close together in space.

9. The NOESY spectrum shows correlations from H$_2$ to H$_1$ / H$_5$ and correlations from H$_3$ to H$_1$ / H$_5$ indicating that the alkene protons H$_2$ and H$_3$ are *trans* to each other across the double bond. If the alkene protons were *cis* to one another, the H$_2$→H$_5$ and H$_3$→H$_1$ correlations would be absent.

^{1}H–^{1}H NOESY spectrum of *trans*-2-octen-4-one (DMSO-*d*$_6$, 400 MHz)

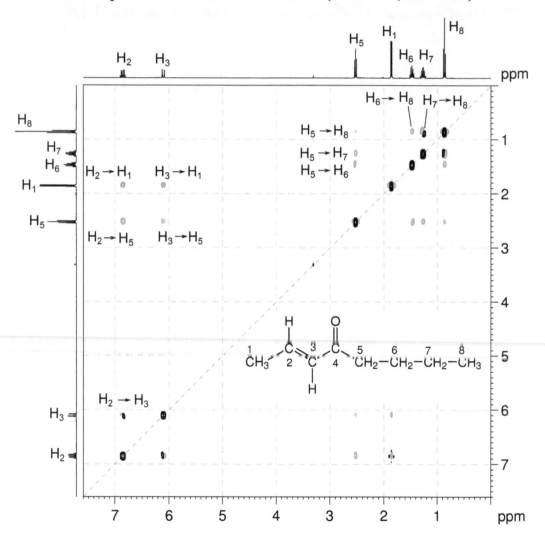

Problem 12

Question:

The ^1H and $^{13}C\{^1H\}$ NMR spectra of 3-nitrobenzaldehyde ($C_7H_5NO_3$) recorded in $CDCl_3$ solution at 298 K and 500 MHz are given below.

The ^1H NMR spectrum has signals at δ 7.82 (H_5), 8.28 (H_6), 8.51 (H_4), 8.73 (H_2) and 10.15 (H_7) ppm.

The $^{13}C\{^1H\}$ NMR spectrum has signals at δ 124.4 (C_2), 128.6 (C_4), 130.5 (C_5), 134.8 (C_6), 137.5 (C_1), 148.8 (C_3) and 189.9 (C_7) ppm.

Also given on the following pages are the ^1H–^1H COSY, ^1H–^{13}C me-HSQC, ^1H–^{13}C HMBC, ^1H–^1H NOESY and INADEQUATE spectra. For each 2D spectrum, indicate which correlation gives rise to each cross-peak by placing an appropriate label in the box provided.

Solution:

3-Nitrobenzaldehyde

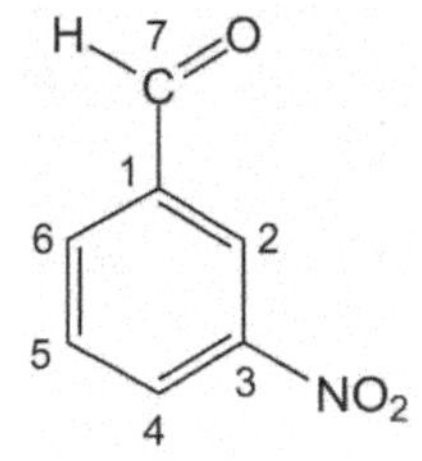

1. ^1H–^1H COSY spectra show coupled sets of protons, and are symmetrical about a diagonal. In this COSY spectrum, there are two $^3J_{H-H}$ correlations above the diagonal ($H_4 \rightarrow H_5$; $H_6 \rightarrow H_5$), and three $^4J_{H-H}$ correlations ($H_2 \rightarrow H_4$; $H_2 \rightarrow H_6$; $H_4 \rightarrow H_6$).

^1H–^1H COSY spectrum of 3-nitrobenzaldehyde (CDCl₃, 500 MHz)

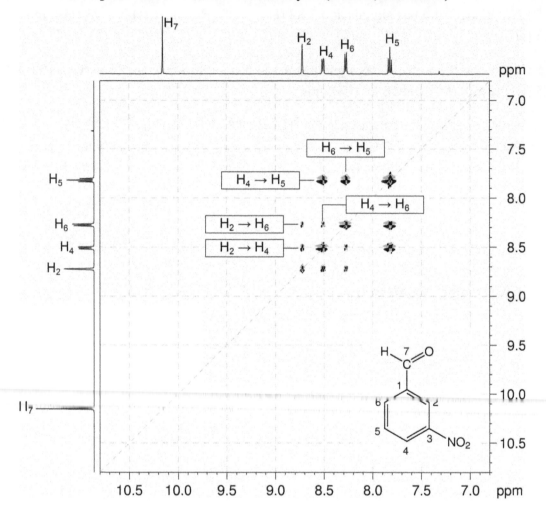

2. The ^{1}H–^{13}C me-HSQC spectrum shows direct (one-bond) correlations between proton and carbon nuclei, so there will be cross-peaks between H$_2$ and C$_2$, H$_4$ and C$_4$, H$_5$ and C$_5$, H$_6$ and C$_6$ and also between H$_7$ and C$_7$. There are no negative (red) cross-peaks in the spectrum as there are no CH$_2$ groups.

^{1}H–^{13}C me-HSQC spectrum of 3-nitrobenzaldehyde (CDCl$_3$, 500 MHz)

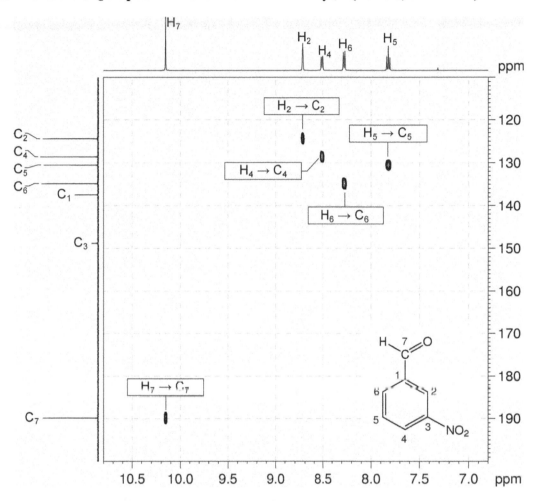

3. In HMBC spectra, remember that, for aromatic systems, the three-bond coupling $^3J_{\text{C–H}}$ is typically the larger long-range coupling and gives rise to the strongest cross-peaks.

^{1}H–^{13}C HMBC spectrum of 3-nitrobenzaldehyde (CDCl$_3$, 500 MHz)

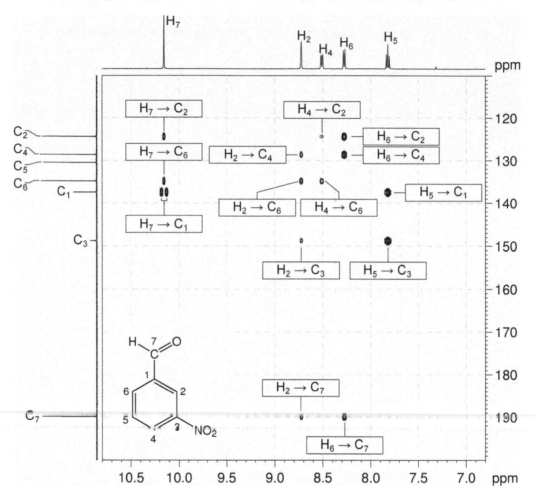

4. H_2 has correlations to C_3, C_4, C_6 and C_7. H_4 correlates to C_2 and C_6. H_5 correlates to C_1 and C_3. H_6 correlates to C_2, C_4 and C_7.

5. H_7 correlates to C_1 (the *ipso* carbon), C_2 and C_6. The correlation between H_7 and C_1 appears as a doublet due to the unusually large $^2J_{\text{C–H}}$ found for aldehydes (~25 Hz).

6. ^1H–^1H NOESY spectra show cross-peaks (off-diagonal peaks) at positions where a proton whose resonance appears on the horizontal axis is close in space to another whose resonance appears on the vertical axis.

7. H_7 correlates to H_2 and H_6, H_4 correlates to H_5 and H_6 correlates to H_5.

^1H–^1H NOESY spectrum of 3-nitrobenzaldehyde (DMSO-d_6, 400 MHz)

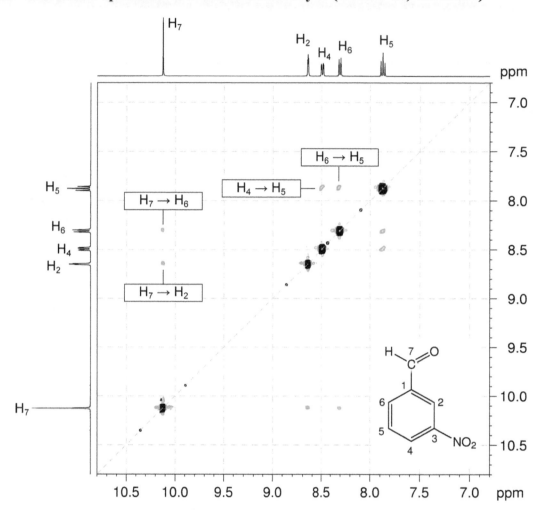

8. The INADEQUATE spectrum shows direct $^{13}C-^{13}C$ connectivity. There are correlations between C_1 and C_2, C_2 and C_3, C_3 and C_4, C_4 and C_5, C_5 and C_6, C_1 and C_6, and also between C_1 and C_7.

INADEQUATE spectrum of 3-nitrobenzaldehyde (CDCl₃, 150 MHz)

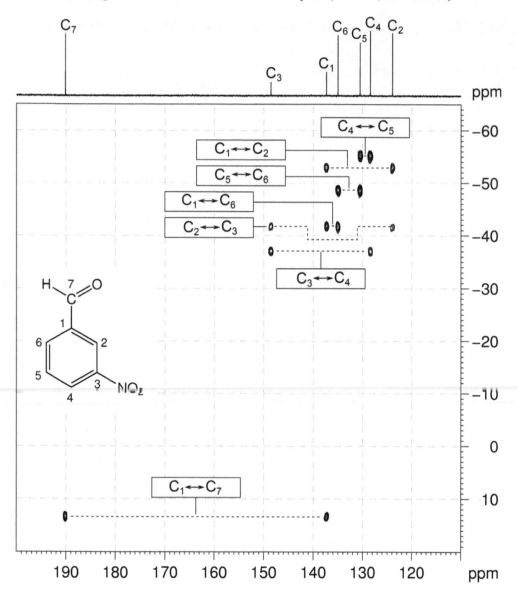

Question:

The ^{1}H and ^{13}C{^{1}H} NMR spectra of 3-iodotoluene (C$_7$H$_6$I) recorded in CDCl$_3$ solution at 298 K and 600 MHz are given below.

The ^{1}H NMR spectrum has signals at δ 2.28 (H$_7$), 6.96 (H$_5$), 7.11 (H$_6$), 7.48 (H$_4$) and 7.53 (H$_2$) ppm.

The ^{13}C{^{1}H} NMR spectrum has signals at δ 21.0, 94.3, 128.3, 129.9, 134.4, 138.0 and 140.2 ppm.

Use the me-HSQC spectrum to assign the protonated carbon signals, and then use this information to produce a schematic HMBC spectrum, showing where all of the cross-peaks would be.

Solution:

3-Iodotoluene

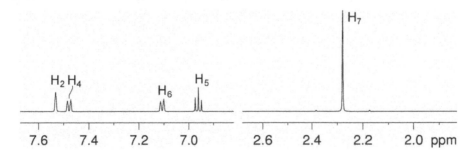

1. The assignments for the ^{1}H NMR spectrum are given.

^{1}H NMR spectrum of 3-iodotoluene (CDCl$_3$, 600 MHz)

2. The $^1H-^{13}C$ me-HSQC easily identifies the protonated carbon resonances: C_2 at 138.0, C_4 at 134.4, C_5 at 129.9, C_6 at 128.3 and C_7 at 21.0 ppm.

3. The non-protonated carbon resonances are therefore those at 94.3 and 140.2 ppm. Based on the chemical shifts, C_3 is the high-field resonance (94.3 ppm) and C_1 the low-field resonance (140.2 ppm).

$^1H-^{13}C$ me-HSQC spectrum of 3-iodotoluene (CDCl$_3$, 600 MHz)

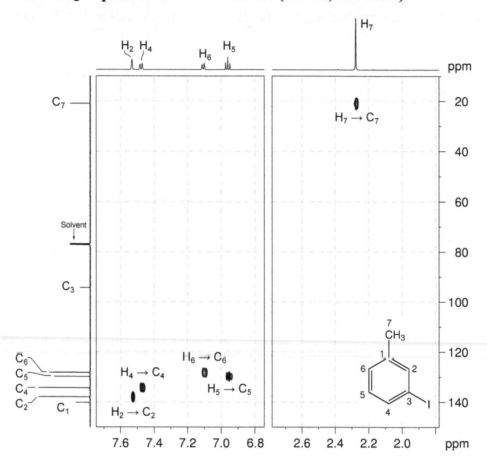

4. Remember that, in aromatic systems, the three-bond coupling $^3J_{C-H}$ is typically the larger long-range coupling and gives rise to the strongest cross-peaks. Benzylic protons typically correlate to the *ipso* carbon (two-bond correlation) and the *ortho* carbons (three-bond correlations).

5. Beginning with the methyl protons (H_7) we would expect correlations to the *ipso* carbon (C_1) and the two carbon atoms *ortho* to the methyl substituent – C_2 and C_6.

6. H_2 will correlate to the carbon nuclei three bonds removed (*i.e. meta*): C_4 and C_6. There is also a three-bond correlation to C_7 (the benzylic carbon).

7. H_4 will correlate to the *meta* carbon nuclei: C_2 and C_6.

8. H_5 will correlate to the *meta* carbon nuclei: C_1 and C_3.

9. H_6 will correlate to the *meta* carbon nuclei: C_2 and C_4. There is also a three-bond correlation to C_7 (the benzylic carbon).

Predicted ^1H–^{13}C HMBC spectrum of 3-iodotoluene (CDCl$_3$, 600 MHz)

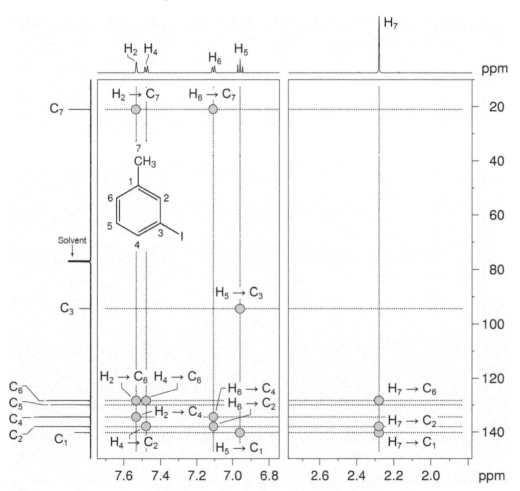

Problem 14

Question:

The ^{1}H and ^{13}C{^{1}H} NMR spectra of 8-hydroxy-5-nitroquinoline ($C_9H_6N_2O_3$) recorded in DMSO-d_6 solution at 298 K and 400 MHz are given below.

The ^{1}H NMR spectrum has signals at δ 7.14 (H$_7$), 7.82 (H$_3$), 8.48 (H$_6$), 8.97 (H$_2$) and 9.08 (H$_4$) ppm. The hydroxyl proton is not shown.

The ^{13}C{^{1}H} NMR spectrum has signals at δ 110.0 (C$_7$), 122.5 (C$_{10}$), 125.2 (C$_3$), 129.1 (C$_6$), 132.4 (C$_4$), 135.0 (C$_5$), 137.2 (C$_9$), 149.1 (C$_2$) and 160.7 (C$_8$) ppm.

Also given on the following pages are the ^{1}H–^{1}H COSY, ^{1}H–^{13}C me-HSQC, ^{1}H–^{13}C HMBC and INADEQUATE spectra. For each 2D spectrum, indicate which correlation gives rise to each cross-peak by placing an appropriate label in the box provided.

Solution:

8-Hydroxy-5-nitroquinoline

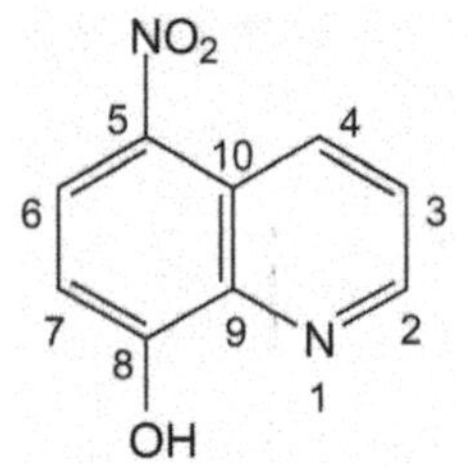

1. ^{1}H–^{1}H COSY spectra show coupled sets of protons, and are symmetrical about a diagonal. In this COSY spectrum, there are three $^{3}J_{H-H}$ correlations above the diagonal ($H_2 \rightarrow H_3$; $H_4 \rightarrow H_3$; $H_6 \rightarrow H_7$), and one $^{4}J_{H-H}$ between H_2 and H_4.

^{1}H–^{1}H COSY spectrum of 8-hydroxy-5-nitroquinoline (DMSO-d_6, 400 MHz)

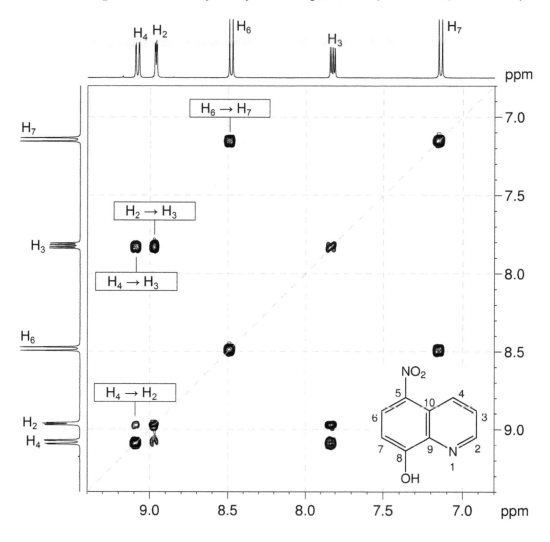

2. The ^1H–^{13}C me-HSQC spectrum shows direct (one-bond) correlations between proton and carbon nuclei, so there will be cross-peaks between H_2 and C_2, H_3 and C_3, H_4 and C_4, H_6 and C_6 and also between H_7 and C_7.

^1H–^{13}C me-HSQC spectrum of 8-hydroxy-5-nitroquinoline (DMSO-d_6, 400 MHz)

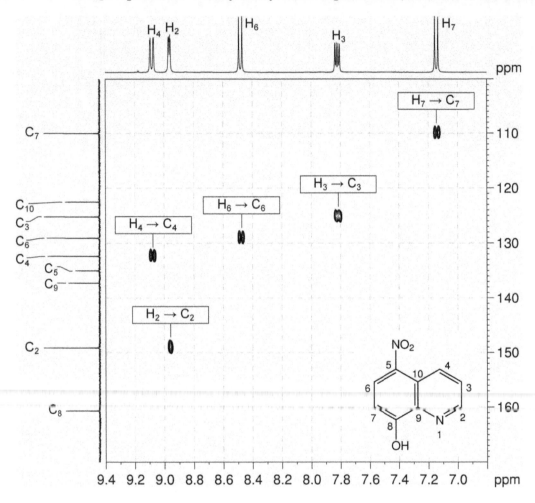

3. In HMBC spectra, remember that, for aromatic systems, the three-bond coupling $^3J_{C-H}$ is typically the larger long-range coupling and gives rise to the strongest cross-peaks.

4. H_2 has correlations to C_3, C_4 and C_9. H_3 correlates to C_2 and C_{10}. H_4 correlates to C_2, C_5 and C_9.

5. H_6 correlates to C_5, C_8 and C_{10}. H_7 correlates to C_5, C_8 and C_9.

^1H–^{13}C HMBC spectrum of 8-hydroxy-5-nitroquinoline (DMSO-d_6, 400 MHz)

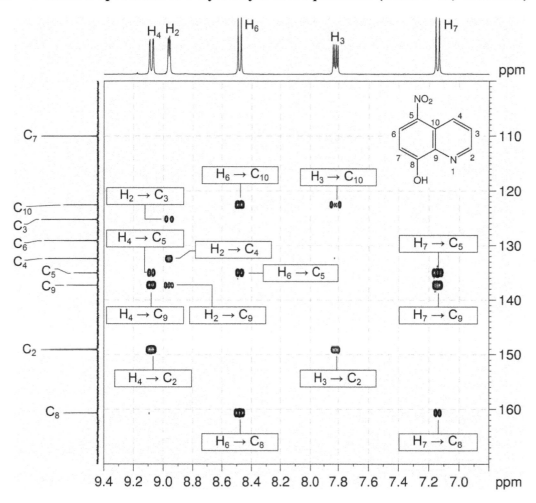

6. The INADEQUATE spectrum shows direct $^{13}C-^{13}C$ connectivity. There are correlations between C_2 and C_3, C_3 and C_4, C_4 and C_{10}, C_5 and C_{10}, C_5 and C_6, C_6 and C_7, C_7 and C_8, C_8 and C_9 and also between C_9 and C_{10}.

INADEQUATE spectrum of 8-hydroxy-5-nitroquinoline (DMSO-d_6, 151 MHz)

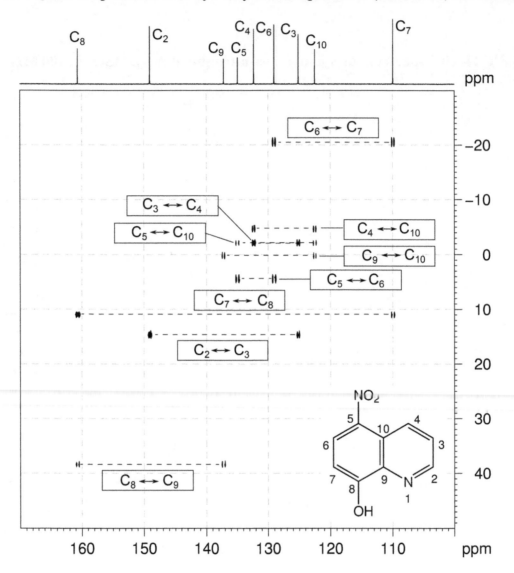

Question:

Identify the following compound.

Molecular Formula: C_6H_6BrN

Solution:

2-Bromo-3-methylpyridine

(2-Bromo-3-picoline)

1. The molecular formula is C_6H_6BrN. Calculate the degree of unsaturation from the molecular formula: replace Br with H, ignore the N atom and remove one H to give an effective molecular formula of C_6H_6 (C_nH_m) which gives the degree of unsaturation as $(n - m/2 + 1) = 6 - 3 + 1 = 4$. The compound contains a combined total of four rings and/or π bonds.

2. The ^{1}H NMR spectrum shows a methyl group (at 2.38 ppm), and three aromatic signals. This accounts for all the hydrogen atoms in the molecular formula and also means that the aromatic ring must contain the N. The chemical shift of one of the aromatic signals (8.19 ppm, H_6) places it adjacent to a nitrogen atom in a pyridine ring.

3. The large coupling constants for each aromatic signal (7.5 and 4.9 Hz) indicate that the three protons are immediately adjacent to one another. The signal at 7.17 ppm is a doublet of doublets indicating it belongs to the aromatic proton located between the other two protons (H_5). The remaining signal at 7.52 ppm must be due to H_4.

^{1}H NMR spectrum of 2-bromo-3-picoline (CDCl$_3$, 600 MHz)

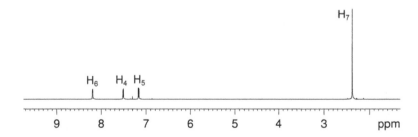

^{1}H NMR spectrum of 2-bromo-3-picoline – expansion

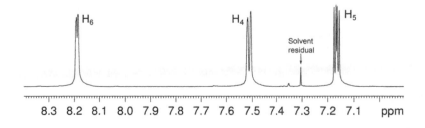

$^{13}C\{^1H\}$ NMR spectrum of 2-bromo-3-picoline (CDCl$_3$, 150 MHz)

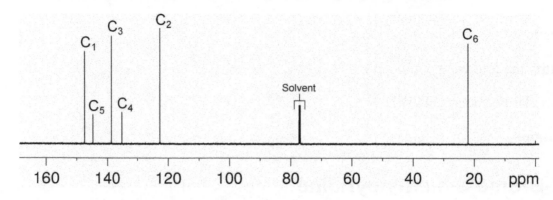

4. The protonated carbon resonances can be identified using the me-HSQC spectrum: C$_6$ at 147.4 ppm, C$_5$ at 122.8 ppm, C$_4$ at 138.7 ppm and the methyl carbon (C$_7$) at 22.0 ppm.

^1H–^{13}C me-HSQC spectrum of 2-bromo-3-picoline (CDCl$_3$, 600 MHz)

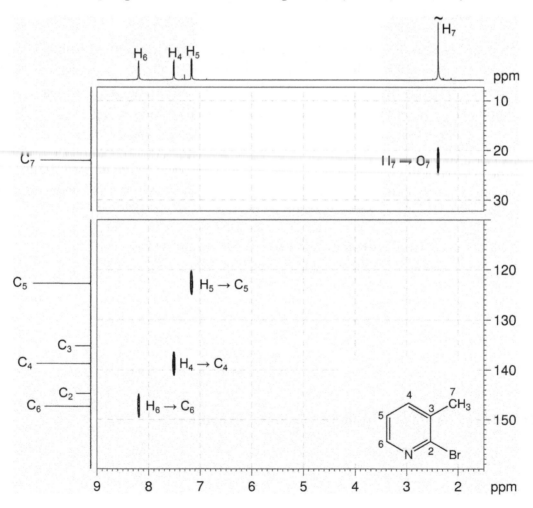

5. There are two possible candidates:

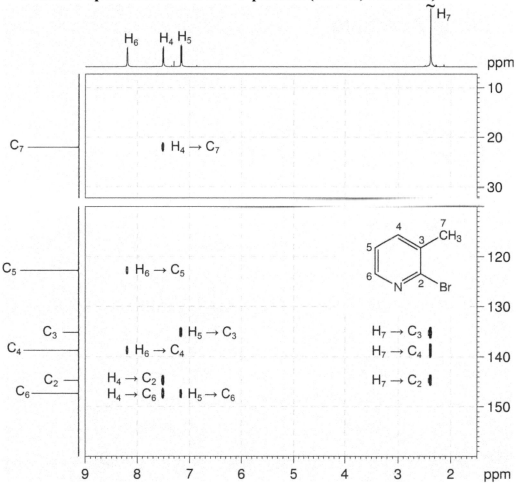

6. The HMBC spectrum can be used to differentiate between the two isomers: there is a strong correlation between H_4 and the methyl group carbon (C_7). In Isomer A, this would be a four-bond correlation, while in Isomer B this would be a three-bond correlation. **Remember** that in aromatic systems, it is the three-bond coupling that is the largest, both around the ring and to benzylic carbon atoms. Isomer A is therefore eliminated as a possibility, and **Isomer B** is identified as the correct structure.

^{1}H–^{13}C HMBC spectrum of 2-bromo-3-picoline (CDCl$_3$, 600 MHz)

7. The non-protonated carbon atoms can also be identified: C_3 at 135.2 ppm ($H_5 \rightarrow C_3$) and C_2 at 144.7 ppm ($H_4 \rightarrow C_2$).

Problem 16

Question:

The ^{1}H NMR spectrum of *trans*-anethole ($C_{10}H_{12}O$) recorded in CDCl$_3$ solution at 298 K and 400 MHz is given below.

The ^{1}H NMR spectrum has signals at δ 1.82 (dd, J = 6.6, 1.7 Hz, 3H, H$_1$), 3.71 (s, 3H, H$_8$), 6.04 (dq, J = 15.8, 6.6 Hz, 1H, H$_2$), 6.30 (dq, J = 15.8, 1.7 Hz, 1H, H$_3$), 6.78 (m, 2H, H$_6$) and 7.21 (m, 2H, H$_5$) ppm.

Use this information to produce schematic diagrams of the COSY and NOESY spectra, showing where all of the cross-peaks and diagonal peaks would be.

Solution:

trans-Anethole

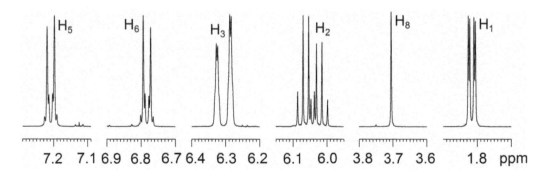

1. The assignments for the ^{1}H NMR spectrum are given.

^{1}H NMR spectrum of *trans*-anethole (CDCl$_3$, 400 MHz)

2. The COSY spectrum has peaks on the diagonal for each unique spin, so the spectrum will contain six diagonal peaks – one at the chemical shift of each resonance in the spectrum.

3. COSY spectra show cross-peaks (off-diagonal peaks) at positions where a proton whose resonance appears on the horizontal axis is directly coupled to another whose resonance appears on the vertical axis.

4. For *trans*-anethole, H_5 and H_6 are coupled to each other and cross-peaks are expected between them.

5. H_2 and H_3 are also coupled to each other and cross-peaks are expected between them.

6. H_1 is coupled to both H_2 and H_3 thus cross-peaks are also expected for H_1–H_2 and H_1–H_3. The H_1–H_3 coupling constant is small and the cross-peaks may appear weaker in intensity.

7. Remember that a COSY spectrum is symmetrical about the diagonal, so the cross-peaks on one side of the diagonal must be reflected on the other side of the diagonal.

Predicted ^1H–^1H COSY spectrum of *trans*-anethole (CDCl$_3$, 400 MHz)

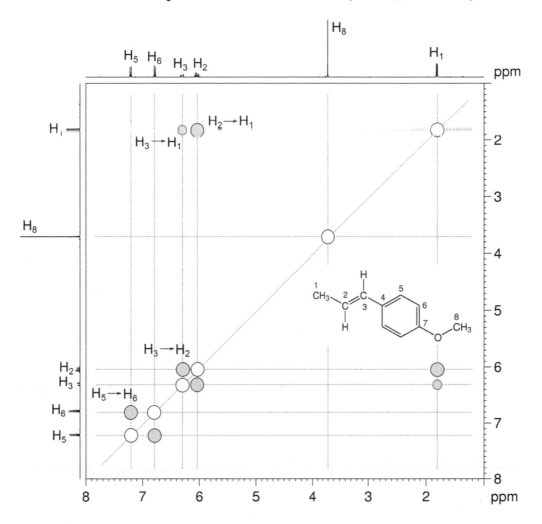

8. The NOESY spectrum has peaks on the diagonal for each unique spin, so the spectrum will contain six diagonal peaks.

9. NOESY spectra show cross-peaks (off-diagonal peaks) at positions where a proton whose resonance appears on the horizontal axis is close in space to another whose resonance appears on the vertical axis.

10. The NOESY spectrum is expected to have correlations for $H_6 \rightarrow H_8$, $H_5 \rightarrow H_6$, $H_5 \rightarrow H_3$, $H_5 \rightarrow H_2$, $H_3 \rightarrow H_2$, $H_3 \rightarrow H_1$ and $H_2 \rightarrow H_1$ above the diagonal.

11. Remember that a NOESY spectrum is symmetrical about the diagonal, so the cross-peaks on one side of the diagonal must be reflected on the other side of the diagonal.

Predicted $^1H-^1H$ NOESY spectrum of *trans*-anethole (CDCl₃, 400 MHz)

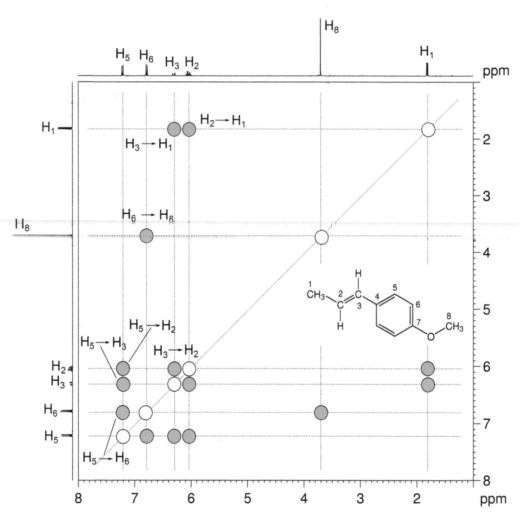

Question:

Identify the following compound.

Molecular Formula: C_5H_{10}

Solution:

cis-2-Pentene

(structure of *cis*-2-pentene showing:)

$$H - C_2 = C_3 - H$$
with CH_3 (position 1) and CH_2-CH_3 (positions 4, 5)

1. The molecular formula is C_5H_{10}. Calculate the degree of unsaturation: the effective molecular formula is C_5H_{10} (C_nH_m) which gives the degree of unsaturation as $(n - m/2 + 1) = 5 - 5 + 1 = 1$. The compound contains one ring or one functional group containing a double bond.

2. The ^1H NMR and me-HSQC spectra establish the presence of two CH_3 groups, one CH_2 group and two alkene $=CH$ groups. The alkene group accounts for all of the degrees of unsaturation, so the compound contains no additional rings or multiple bonds.

^1H NMR spectrum of *cis*-2-pentene (CDCl₃, 400 MHz)

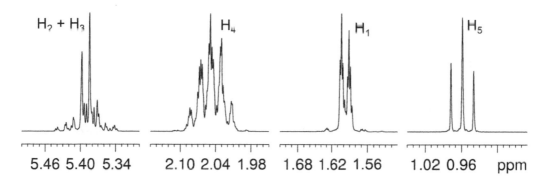

$H_2 + H_3$ · H_4 · H_1 · H_5

5.46 5.40 5.34 · 2.10 2.04 1.98 · 1.68 1.62 1.56 · 1.02 0.96 · ppm

$^{13}C\{^1H\}$ NMR spectrum of *cis*-2-pentene (CDCl₃, 100 MHz)

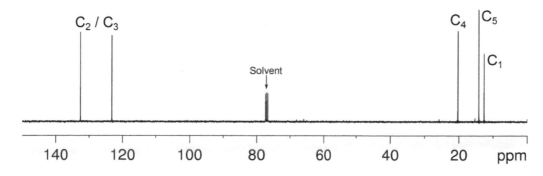

C_2 / C_3 · Solvent · C_4 · C_5 · C_1

140   120   100   80   60   40   20   ppm

3. In the COSY spectrum, the H_4–H_5 and H_3–H_4 correlations afford the =CHCH$_2$CH$_3$ fragment. The H_1–H_2 correlation affords the CH$_3$–CH= fragment. Combining the two fragments affords CH$_3$–CH=CH–CH$_2$CH$_3$.

^{1}H–^{1}H COSY spectrum of *cis*-2-pentene (CDCl$_3$, 400 MHz)

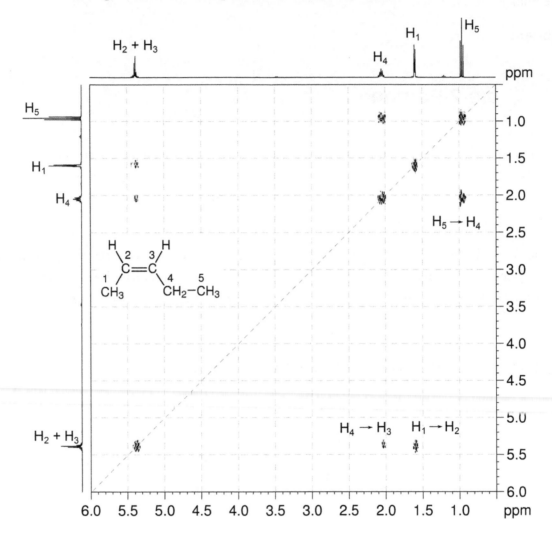

4. The resonances for C_1, C_4 and C_5 can be assigned using the me-HSQC spectrum. Note that although C_2 and C_3 cannot be assigned based on the information provided, the structure of the molecule can still be solved.

^1H–^{13}C me-HSQC spectrum of *cis*-2-pentene (CDCl$_3$, 400 MHz)

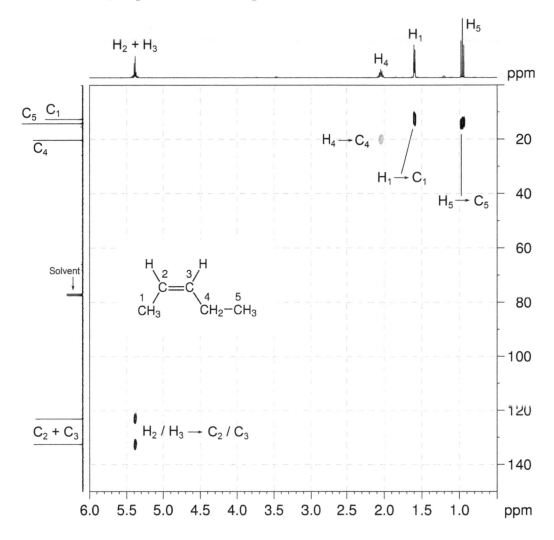

5. The geometry of the double bond can be deduced from the NOESY spectrum. In particular the H₄–H₁ correlation places the CH₂ (H₄) and CH₃ (H₁) groups on the same side of the double bond and thus gives a *cis* geometry for the molecule.

¹H–¹H NOESY spectrum of *cis*-2-pentene (DMSO-*d₆*, 400 MHz)

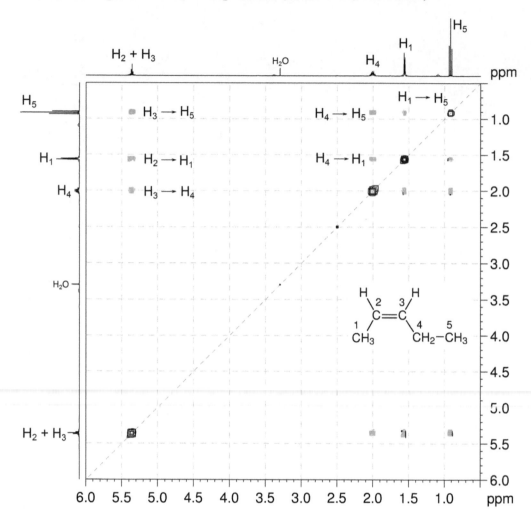

Question:

Identify the following compound.

Molecular Formula: $C_{14}H_{12}O_2$

IR: 1720 cm^{-1}

Solution:

p-Tolyl benzoate

1. The molecular formula is $C_{14}H_{12}O_2$. Calculate the degree of unsaturation from the molecular formula: ignore the O atoms to give an effective molecular formula of $C_{14}H_{12}$ (C_nH_m) which gives the degree of unsaturation as $(n - m/2 + 1) = 14 - 6 + 1 = 9$. The compound contains a combined total of nine rings and/or π bonds.

2. The ^{1}H NMR spectrum indicates that the compound has nine aromatic protons and three aliphatic protons.

3. The presence of nine aromatic protons suggests two aromatic rings.

4. The three aliphatic protons are equivalent and have no visible splitting so these must be an isolated methyl group.

5. IR and 1D NMR spectra establish that the compound is an ester (quaternary ^{13}C resonance at 165.3 ppm).

6. The presence of two aromatic rings and an ester carbonyl satisfies the degree of unsaturation, so there are no more functional groups with double bonds or rings.

7. The methyl resonance in the ^{1}H and ^{13}C spectra (at 2.36 and 20.9 ppm, respectively) eliminates the possibility of a methyl ester, since for a methoxy group both the ^{1}H and ^{13}C signals would appear significantly further downfield.

8. The ^{1}H NMR spectrum shows five unique signals in the aromatic region.

^{1}H NMR spectrum of *p*-tolyl benzoate (CDCl$_3$, 500 MHz)

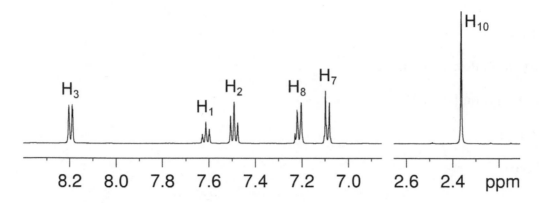

^{13}C{^{1}H} NMR spectrum of *p*-tolyl benzoate (CDCl$_3$, 125 MHz)

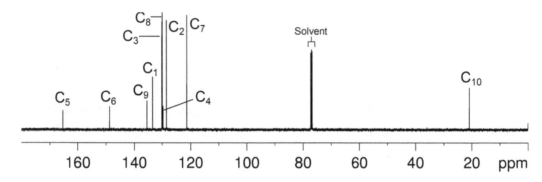

^{13}C{^{1}H} NMR spectrum of *p*-tolyl benzoate – expansion

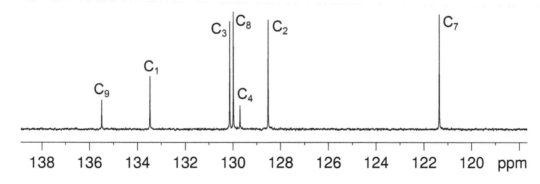

9. The ^{1}H–^{1}H COSY spectrum shows that the resonances at 7.21 and 7.09 ppm are coupled. The coupling pattern of these two signals is consistent with a *para*-disubstituted benzene. The five protons in the three remaining signals (at 8.20, 7.61 and 7.49 ppm for H$_3$, H$_1$ and H$_2$, respectively) constitute a separate, single spin system. The coupling pattern of these three signals is consistent with a mono-substituted benzene ring.

^{1}H–^{1}H COSY spectrum of *p*-tolyl benzoate – expansion A (CDCl$_3$, 500 MHz)

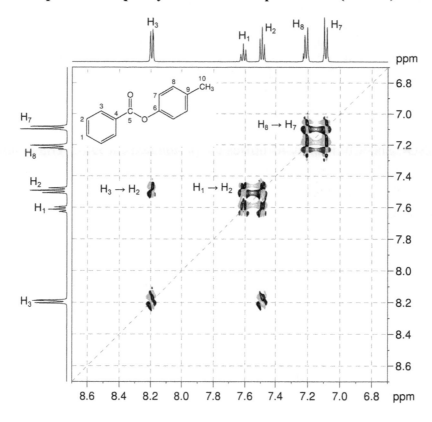

^{1}H–^{1}H COSY spectrum of *p*-tolyl benzoate – expansion B

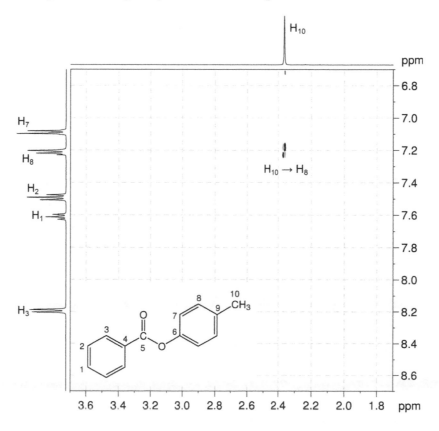

10. There are three possible isomers:

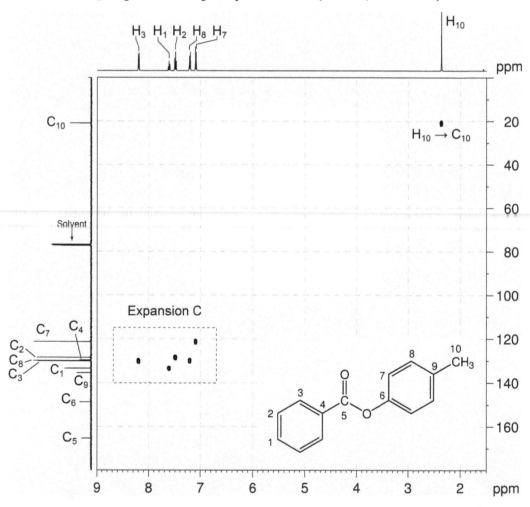

A   B   C

11. The me-HSQC spectrum easily identifies the protonated carbons. The aromatic carbons of the mono-substituted ring – C_1, C_2 and C_3 – are at 133.5, 128.5 and 130.1 ppm, respectively. The protonated carbons of the *para*-disubstituted ring – C_7 and C_8 – are at 121.4 and 130.0 ppm, respectively.

^1H–^{13}C me-HSQC spectrum of *p*-tolyl benzoate (CDCl$_3$, 500 MHz)

^{1}H–^{13}C me-HSQC spectrum of *p*-tolyl benzoate – expansion C

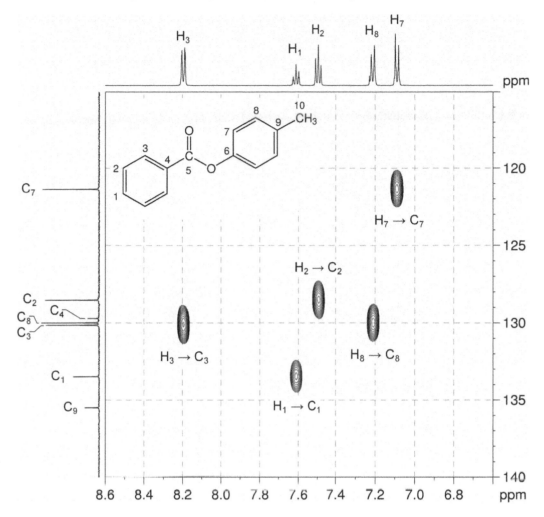

12. In the ^{1}H–^{1}H COSY spectrum there is a correlation between the methyl protons (H_{10}) and the aromatic H_8 protons, identifying the methyl group as one of the substituents on the *para*-disubstituted benzene ring. Isomer **A** may therefore be eliminated.

13. In the HMBC spectrum, there is a correlation between H_3 and the ester carbonyl carbon (at 165.3 ppm).

14. H_3 must be the two protons *ortho* to the substituent of the mono-substituted aromatic ring. In isomer **B**, these protons are four-bonds removed from the carbonyl-carbon, while in isomer **C**, H_3 is three-bonds removed.

15. **Remember**, that in aromatic systems, the three-bond coupling $^3J_{\text{C–H}}$ is typically the larger long-range coupling and gives rise to the stronger cross-peaks. Isomer **B** may therefore be eliminated, leaving **Isomer C** as the correct answer.

^{1}H–^{13}C HMBC spectrum of *p*-tolyl benzoate (CDCl$_3$, 500 MHz)

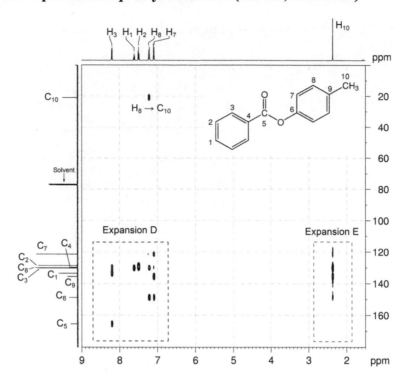

^{1}H–^{13}C HMBC spectrum of *p*-tolyl benzoate – expansion D

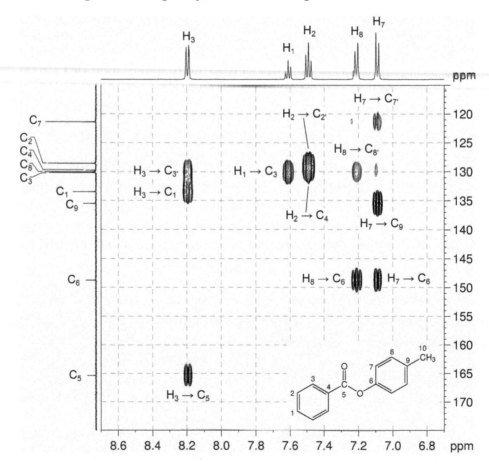

^{1}H–^{13}C HMBC spectrum of *p*-tolyl benzoate – expansion E

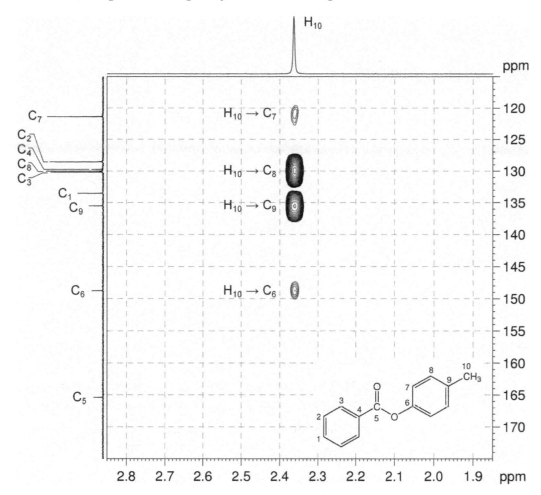

16. In the HMBC spectrum for this compound, there are strong correlations between H_2 and $C_{2'}$, H_3 and $C_{3'}$, H_7 and $C_{7'}$, and H_8 and $C_{8'}$. While these appear to be one-bond correlations, in any substituted aromatic ring in which a mirror plane bisects two of the substituents, these apparent one-bond correlations arise from the $^3J_{C-H}$ interaction of a proton with the carbon which is *meta* to it.

Problem 19

Question:

Identify the following compound.

Molecular Formula: $C_{14}H_{12}O_2$

IR: 1720 cm^{-1}

Solution:

Phenyl *p*-toluate

1. The molecular formula is $C_{14}H_{12}O_2$. Calculate the degree of unsaturation from the molecular formula: ignore the O atoms to give an effective molecular formula of $C_{14}H_{12}$ (C_nH_m) which gives the degree of unsaturation as $(n - m/2 + 1) = 14 - 6 + 1 = 9$. The compound contains a combined total of nine rings and/or π bonds.

2. The ^{1}H NMR spectrum indicates that the compound has nine aromatic protons and three aliphatic protons.

3. The presence of nine aromatic protons suggests two aromatic rings.

4. The three aliphatic protons are equivalent and have no visible splitting so these must be an isolated methyl group.

5. IR and 1D NMR spectra establish that the compound is an ester (quaternary ^{13}C resonance at 165.2 ppm).

6. The presence of two aromatic rings and an ester carbonyl satisfies the degree of unsaturation, so there are no more functional groups with double bonds or rings.

7. The methyl resonance in the ^{1}H and ^{13}C spectra (2.44 and 21.7 ppm, respectively) eliminates the possibility of a methyl ester, since for a methoxy group both the ^{1}H and ^{13}C signals would appear significantly further downfield.

8. The ^{1}H NMR spectrum shows five unique signals in the aromatic region.

^{1}H NMR spectrum of phenyl *p*-toluate (CDCl$_3$, 500 MHz)

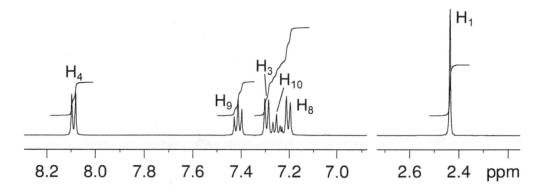

^{13}C{^{1}H} NMR spectrum of phenyl *p*-toluate (CDCl$_3$, 125 MHz)

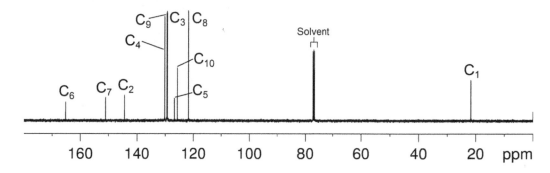

^{13}C{^{1}H} NMR spectrum of phenyl *p*-toluate – expansion

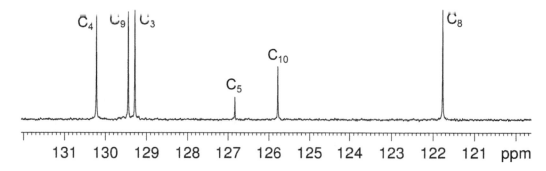

9. The ^{1}H–^{1}H COSY spectrum shows that the resonances at 8.09 and 7.29 ppm are coupled. The coupling pattern of these two signals is consistent with a *para*-disubstituted benzene ring. The three remaining signals (at 7.41, 7.25 and 7.20 ppm for H$_9$, H$_{10}$ and H$_8$, respectively) constitute a separate, single spin system. The coupling patterns of these three signals are consistent with a mono-substituted benzene ring.

10. In the ^{1}H–^{1}H COSY spectrum, there is a correlation between the methyl protons H$_1$ and the aromatic protons H$_3$, identifying the methyl group as one of the substituents on the *para*-disubstituted benzene ring *i.e.* a *p*-tolyl group.

^{1}H–^{1}H COSY spectrum of phenyl p-toluate – expansion A (CDCl$_3$, 500 MHz)

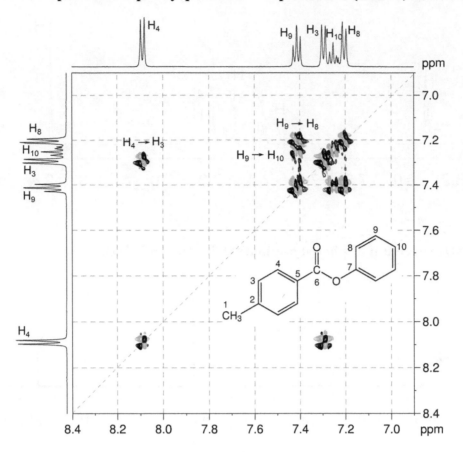

^{1}H–^{1}H COSY spectrum of phenyl p-toluate – expansion B

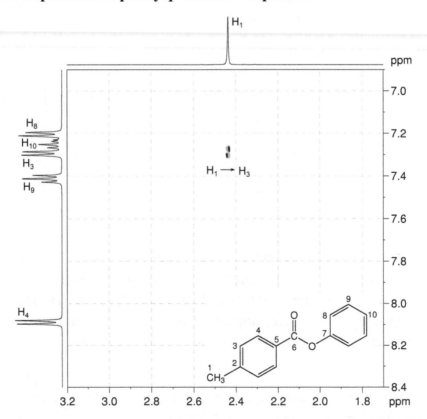

11. The me-HSQC spectrum easily identifies the protonated carbons. The aromatic carbons of the *para*-substituted ring – C₃ and C₄ – are at 129.3 and 130.2 ppm, respectively. The aromatic carbons of the mono-substituted ring – C₈, C₉ and C₁₀ – are 121.8, 129.5 and 125.8 ppm, respectively.

¹H–¹³C me-HSQC spectrum of phenyl *p*-toluate (CDCl₃, 500 MHz)

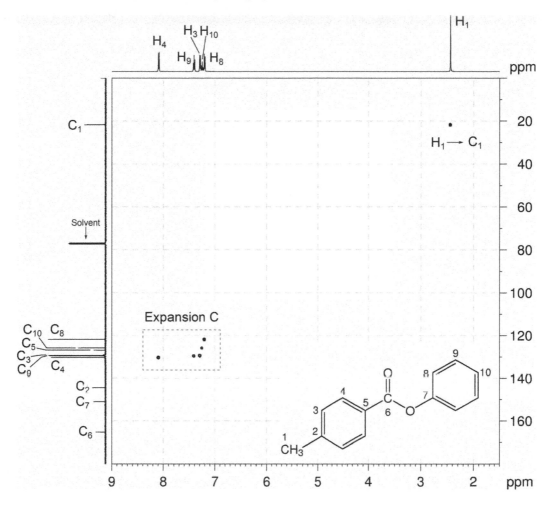

$^1H-^{13}C$ me-HSQC spectrum of phenyl p-toluate – expansion C

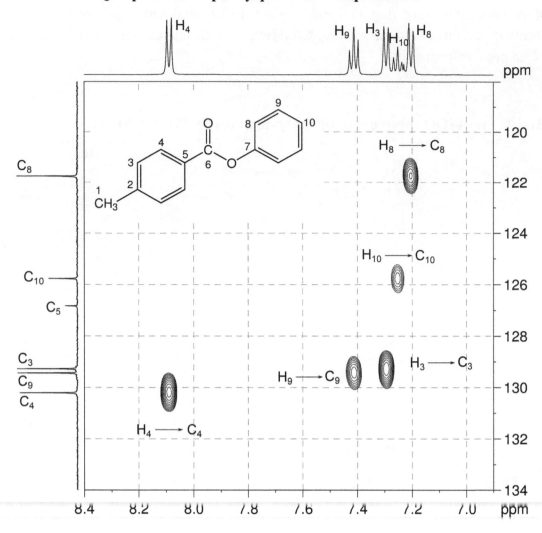

12. In the HMBC spectrum, there is a correlation between H$_4$ and the ester carbonyl carbon C$_6$. H$_4$ belongs to the *p*-tolyl group and this indicates the tolyl group is bound directly to the carbonyl carbon leaving the mono-substituted benzene ring to be attached to the oxygen atom of the ester group.

^{1}H–^{13}C HMBC spectrum of phenyl *p*-toluate – expansion D (CDCl$_3$, 500 MHz)

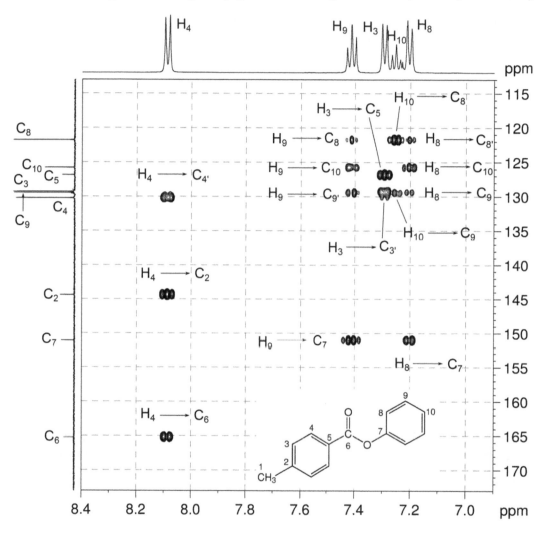

^{1}H–^{13}C HMBC spectrum of phenyl *p*-toluate – expansion E

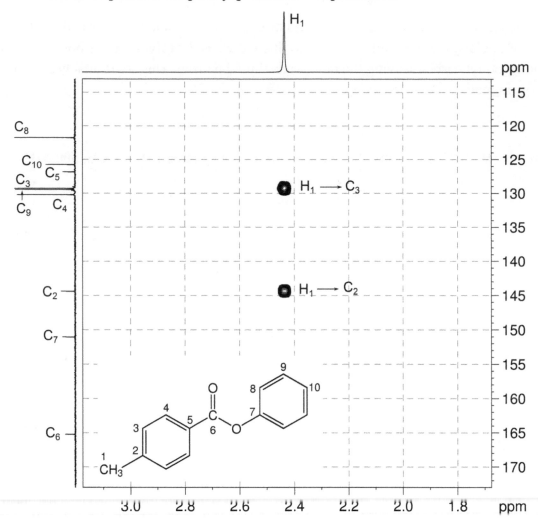

13. In the HMBC spectrum for this compound, there are strong correlations between H$_3$ and C$_{3'}$, H$_4$ and C$_{4'}$, H$_8$ and C$_{8'}$ and also between H$_9$ and C$_{9'}$. While these appear to be one-bond correlations, in any substituted aromatic ring in which a mirror plane bisects two of the substituents, these apparent one-bond correlations arise from the $^3J_{C-H}$ interaction of a proton with the carbon which is *meta* to it.

Question:

Identify the following compound.

Molecular Formula: $C_{14}H_{12}O_2$

IR: 1751 cm^{-1}

Solution:

4-Biphenylyl acetate

1. The molecular formula is $C_{14}H_{12}O_2$. Calculate the degree of unsaturation from the molecular formula: ignore the O atoms to give an effective molecular formula of $C_{14}H_{12}$ (C_nH_m) which gives the degree of unsaturation as $(n - m/2 + 1) = 14 - 6 + 1 = 9$. The compound contains a combined total of nine rings and/or π bonds.

2. The ^{1}H NMR spectrum indicates that the compound has nine aromatic protons and three aliphatic protons.

3. The presence of nine aromatic protons suggests two aromatic rings.

4. The three aliphatic protons are equivalent and have no visible splitting so these must be an isolated methyl group.

5. IR and 1D NMR spectra establish that the compound is an ester (quaternary ^{13}C resonance at 168.9 ppm).

6. The presence of two aromatic rings and an ester carbonyl satisfies the degree of unsaturation, so there are no more functional groups with double bonds or rings.

7. The methyl resonance in the ^{1}H and ^{13}C spectra (at 2.30 and 20.2 ppm, respectively) eliminates the possibility of a methyl ester, since for a methoxy group both the ^{1}H and ^{13}C signals would appear significantly further downfield.

8. The ^{1}H NMR spectrum shows three unique signals and two overlapping signals in the aromatic region.

^{1}H NMR spectrum of 4-biphenylyl acetate (CDCl$_3$, 500 MHz)

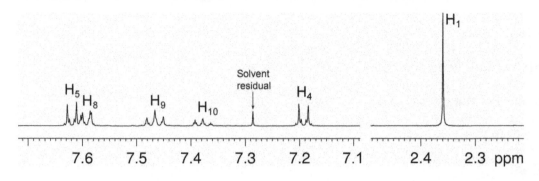

^{13}C{^{1}H} NMR spectrum of 4-biphenylyl acetate (CDCl$_3$, 100 MHz)

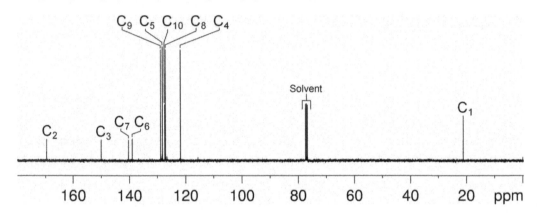

^{13}C{^{1}H} NMR spectrum of 4-biphenylyl acetate — expansion

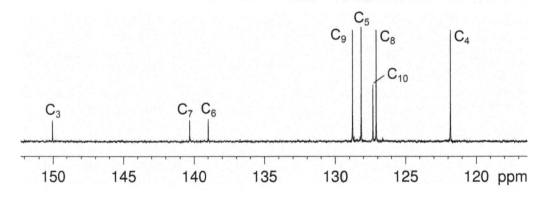

9. The ^1H–^1H COSY spectrum shows that the resonances at 7.57 and 7.15 ppm are coupled. The coupling pattern of these two signals is consistent with a *para*-disubstituted benzene ring. The three remaining signals (at 7.55, 7.42 and 7.33 ppm for H_8, H_9 and H_{10}, respectively) constitute a separate, single spin system. The integrations and coupling patterns of these three signals are consistent with a mono-substituted benzene ring.

^1H–^1H COSY spectrum of 4-biphenylyl acetate – expansion A (CDCl₃, 400 MHz)

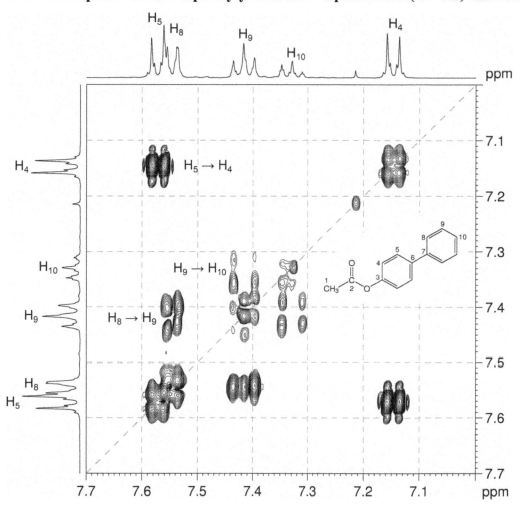

10. The me-HSQC spectrum easily identifies the protonated carbons. The aromatic carbons of the *para*-disubstituted ring – C_4 and C_5 – are at 121.8 and 128.2 ppm, respectively. The aromatic carbons of the mono-substituted ring – C_8, C_9 and C_{10} – are 127.1, 128.8 and 127.3 ppm, respectively.

$^1H–^{13}C$ me-HSQC spectrum of 4-biphenylyl acetate – expansion B (CDCl₃, 400 MHz)

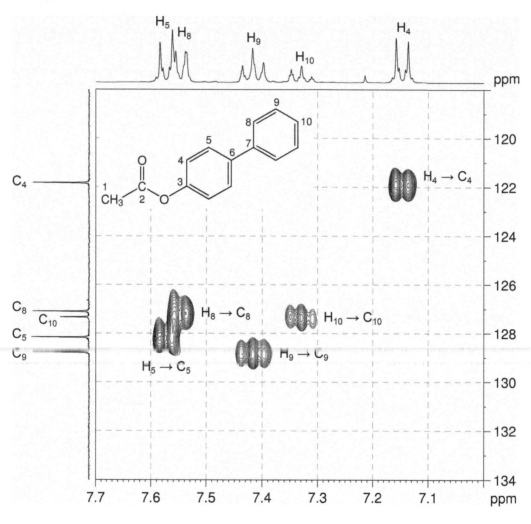

11. In the HMBC spectrum, there is a correlation between the methyl protons H_1 and the ester carbonyl carbon C_2. The methyl group is bound directly to the carbonyl carbon.

12. Note that the one-bond coupling between H_1 and C_1 is visible in the HMBC spectrum as a large doublet.

13. **Remember**, that in aromatic systems, the three-bond coupling $^3J_{C-H}$ is typically the larger long-range coupling and gives rise to the stronger cross-peaks. The non-protonated carbons can thus be assigned (C_3 at 150.1, C_6 at 139.0 and C_7 at 140.4 ppm).

14. Note that there is a strong two-bond correlation between H_4 and C_3 typical for O-bound carbon atoms.

15. The inter-ring correlations H_5 to C_7 and H_8 to C_6 indicate the presence of the biphenyl group.

^1H–^{13}C HMBC spectrum of 4-biphenylyl acetate (CDCl$_3$, 400 MHz)

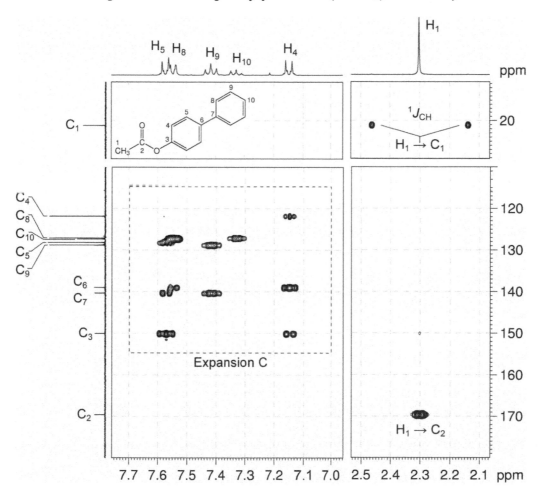

^1H–^{13}C HMBC spectrum of 4-biphenylyl acetate – expansion C

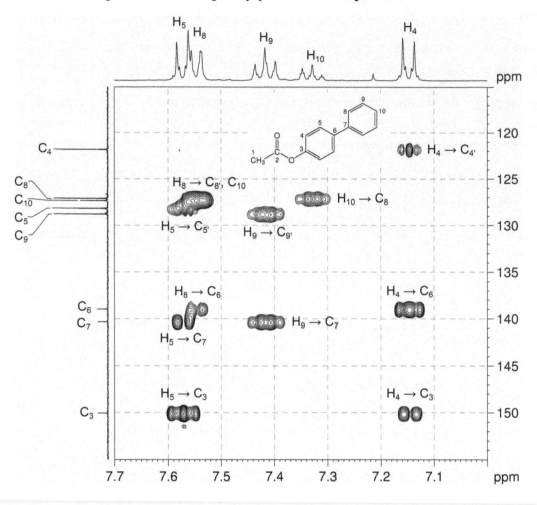

16. In the HMBC spectrum for this compound, there are strong correlations between H_4 and $C_{4'}$, H_5 and $C_{5'}$, H_8 and $C_{8'}$ and also between H_9 and $C_{9'}$. While these appear to be one-bond correlations, in any substituted aromatic ring in which a mirror plane bisects two of the substituents, these apparent one-bond correlations arise from the $^3J_{C-H}$ interaction of a proton with the carbon which is *meta* to it.

Question:

Identify the following compound.

Molecular Formula: $C_{14}H_{12}O_2$

IR: 1678 cm^{-1}

Solution:

4'-Phenoxyacetophenone

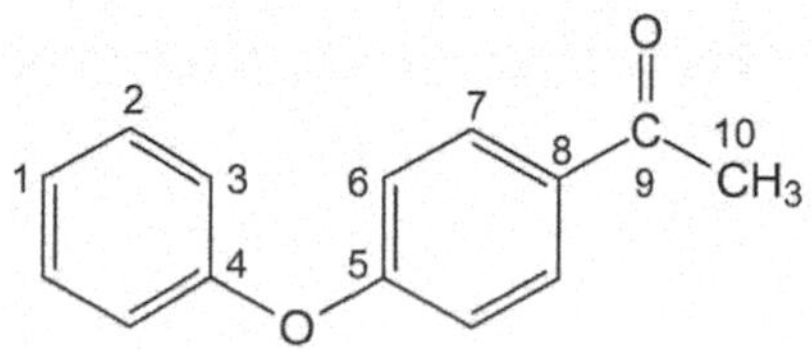

1. The molecular formula is $C_{14}H_{12}O_2$. Calculate the degree of unsaturation from the molecular formula: ignore the O atoms to give an effective molecular formula of $C_{14}H_{12}$ (C_nH_m) which gives the degree of unsaturation as $(n - m/2 + 1) = 14 - 6 + 1 = 9$. The compound contains a combined total of nine rings and/or π bonds.

2. The ^1H NMR spectrum indicates that the compound has nine aromatic protons and three aliphatic protons.

3. The presence of nine aromatic protons suggests two aromatic rings.

4. The three aliphatic protons are equivalent and have no visible splitting so these must be an isolated methyl group.

5. IR and 1D NMR spectra establish the presence of a ketone (^{13}C resonance at 196.7 ppm).

6. The presence of two aromatic rings and a ketone satisfies the degree of unsaturation, so there are no more functional groups with double bonds or rings.

7. The methyl resonance in the ^1H and ^{13}C spectra (at 2.56 and 26.4 ppm, respectively) eliminates the possibility of a methoxy group, since for a methoxy group both the ^1H and ^{13}C signals would appear significantly further downfield.

8. The ^1H NMR spectrum shows five unique signals in the aromatic region.

^{1}H NMR spectrum of 4'-phenoxyacetophenone (CDCl$_3$, 400 MHz)

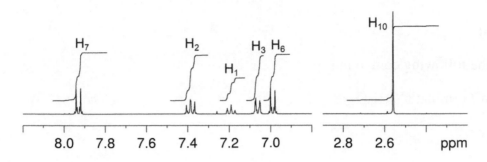

^{13}C{^{1}H} NMR spectrum of 4'-phenoxyacetophenone (CDCl$_3$, 100 MHz)

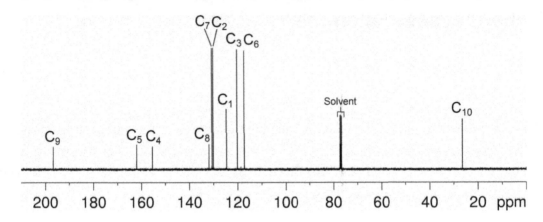

9. The ^1H–^1H COSY spectrum shows that the resonances at 7.93 and 6.99 ppm are coupled. The coupling pattern of these two signals is consistent with a *para*-disubstituted benzene ring.

10. The three remaining signals (at 7.39, 7.19 and 7.06 ppm for H_2, H_1 and H_3, respectively) constitute a separate, single spin system. The coupling patterns of these three signals are consistent with a mono-substituted benzene ring.

^1H–^1H COSY spectrum of 4'-phenoxyacetophenone – expansion A (CDCl$_3$, 400 MHz)

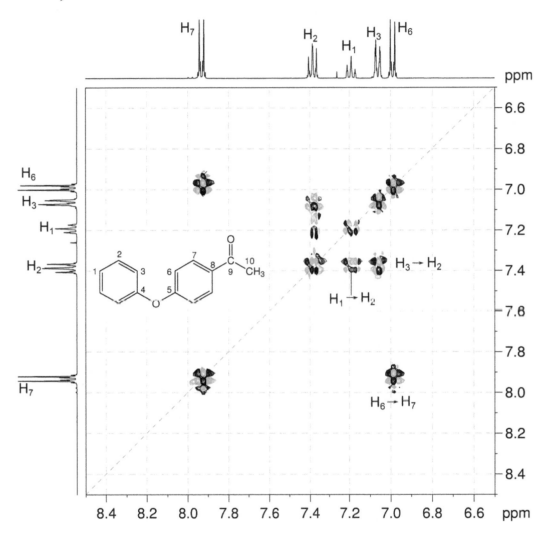

11. The me-HSQC spectrum easily identifies the protonated carbons. The aromatic carbons of the *para*-substituted ring are at 117.3 and 130.6 ppm.

12. The aromatic carbons of the mono-substituted ring – C_1, C_2 and C_3 – are 124.6, 130.0 and 120.1 ppm, respectively.

^1H–^{13}C me-HSQC spectrum of 4'-phenoxyacetophenone (CDCl$_3$, 400 MHz)

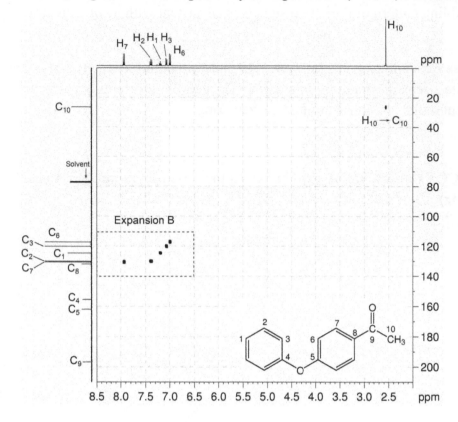

^1H–^{13}C me-HSQC spectrum of 4'-phenoxyacetophenone – expansion B

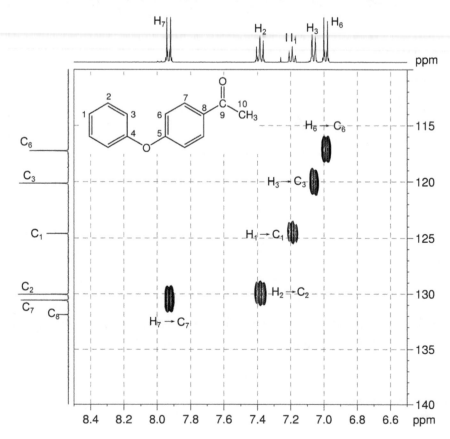

13. In the HMBC spectrum, there is a correlation between the methyl protons H_{10} and the ketone carbonyl carbon C_9. The methyl group must be bound directly to the carbonyl carbon.

14. In the HMBC spectrum, there is also a correlation between one of the proton signals of the *para*-disubstituted benzene ring with the carbonyl carbon (H_7–C_9). The *para*-disubstituted ring must be bound directly to the carbonyl carbon and the ^1H signal at 7.93 ppm is due to the aromatic proton closest to the carbonyl group (H_7).

15. Thus the ^1H signal at 6.99 ppm must be due to H_6. We can now assign C_6 and C_7 to the ^{13}C signals at 117.3 and 130.6 ppm, respectively using the me-HSQC spectrum.

16. In the HMBC spectrum, the correlations from aromatic protons H_2 and H_3 to the ^{13}C resonance at 155.5 ppm identify this signal as that due the quaternary aromatic carbon of the mono-substituted ring (C_4).

^1H–^{13}C HMBC spectrum of 4'-phenoxyacetophenone (CDCl$_3$, 400 MHz)

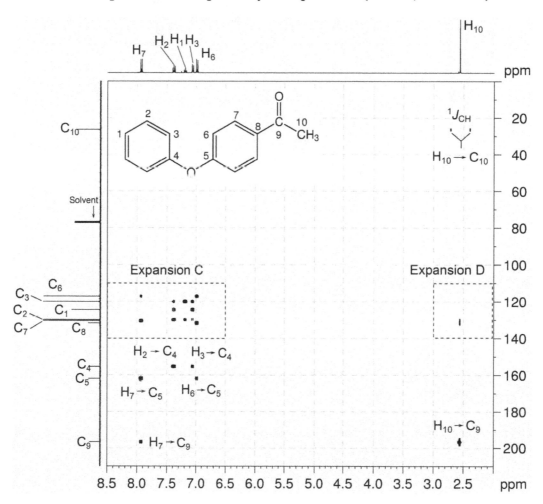

$^1H-^{13}C$ HMBC spectrum of 4'-phenoxyacetophenone – expansion C

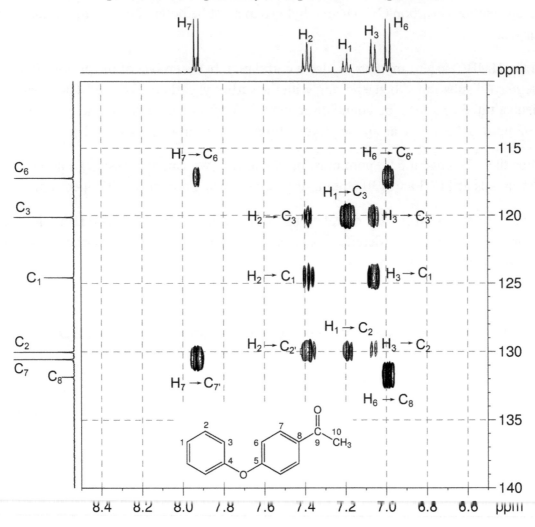

14. In expansion D of the HMBC spectrum, the H_{10}–C_7/C_8 correlation identifies C_8 as the quaternary aromatic carbon directly bonded to the carbonyl carbon.

15. Thus the remaining quaternary aromatic carbon C_5 can be identified by the H_7–C_5 and H_6–C_5 correlations in the HMBC spectrum.

16. The remaining oxygen atom in the molecule must be located between the two aromatic rings and the downfield chemical shifts of C_4 and C_5 are consistent with this formulation.

^1H–^{13}C HMBC spectrum of 4'-phenoxyacetophenone – expansion D

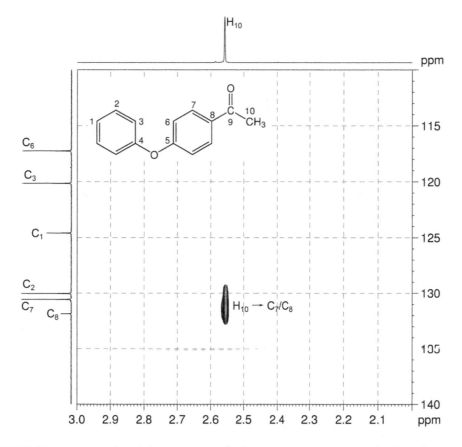

17. In the HMBC spectrum for this compound, there are strong correlations between H_2 and $C_{2'}$, H_3 and $C_{3'}$, H_6 and $C_{6'}$ and also between H_7 and $C_{7'}$. While these appear to be one-bond correlations, in any substituted aromatic ring in which a mirror plane bisects two of the substituents, these apparent one-bond correlations arise from the $^3J_{C–H}$ interaction of a proton with the carbon which is *meta* to it.

Problem 22

Question:

Identify the following compound.

Molecular Formula: $C_{12}H_{16}O$

IR Spectrum: 1686 cm^{-1}

Solution:

4'-*tert*-Butylacetophenone

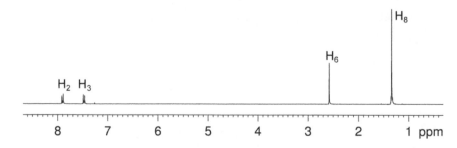

1. The molecular formula is $C_{12}H_{16}O$. Calculate the degree of unsaturation from the molecular formula: ignore the O atom to give an effective molecular formula of $C_{12}H_{16}$ (C_nH_m) which gives the degree of unsaturation as $(n - m/2 + 1) = 12 - 8 + 1 = 5$. The compound contains a combined total of five rings and/or π bonds.

2. The IR and 1D NMR data establish that the compound is aromatic and contains a ketone functional group (quaternary ^{13}C resonance at 197.8 ppm). The aromatic ring and the carbonyl group account for all of the degrees of unsaturation, so the compound contains no additional rings or multiple bonds.

3. There are two aromatic proton resonances in the ^1H NMR spectrum. On the basis of the coupling pattern, the ring is *para*-disubstituted, but one cannot readily distinguish which resonance belongs to H_2 and which to H_3.

4. The ^1H NMR spectrum has one three-proton resonance at 2.58 ppm (H_6), and a nine-proton resonance at 1.34 ppm (H_8). The chemical shift of the resonance at 2.58 ppm is consistent with a methyl group either bound directly to the aromatic ring or to the ketone, while the resonance at 1.34 ppm is due to a *tert*-butyl group.

^1H NMR spectrum of 4'-*tert*-butylacetophenone (CDCl₃, 300 MHz)

$^{13}C\{^1H\}$ NMR spectrum of 4'-*tert*-butylacetophenone (CDCl$_3$, 75 MHz)

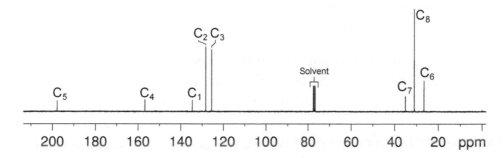

5. There are two possible isomers:

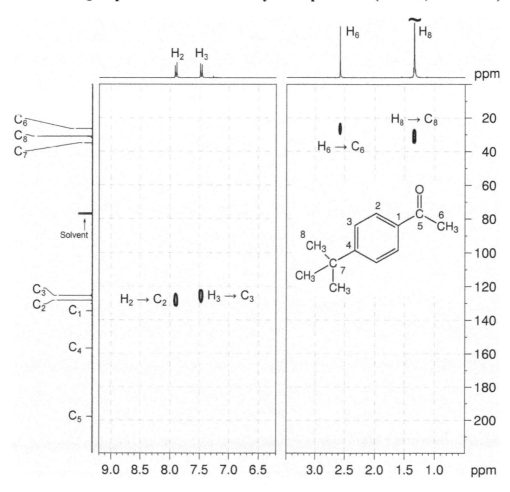

6. The ^1H–^{13}C me-HSQC spectrum easily identifies the protonated carbon resonances: C$_6$ at 26.5 ppm, C$_8$ at 31.0 ppm and the protonated aromatic carbons at 125.5 and 128.3 ppm.

^1H–^{13}C me-HSQC spectrum of 4'-*tert*-butylacetophenone (CDCl$_3$, 300 MHz)

7. The ^1H–^{13}C HMBC spectrum has a correlation between H_6 and the carbonyl carbon C_5 so the methyl group is close to the carbonyl group. This identifies **Isomer A** as the correct structure.

8. The HMBC spectrum can be used to identify and assign the remaining carbon and proton resonances. **Remember** that, in aromatic systems, the three-bond coupling $^3J_{C-H}$ is typically the larger long-range coupling and gives rise to the strongest cross-peaks.

9. The *tert*-butyl protons (H_8) correlate to three resonances in the HMBC spectrum – the first is a strong correlation between H_8 and $C_{8'}$. While this appears to be a one-bond correlation, in *tert*-butyl groups, isopropyl groups or compounds with a *gem*-dimethyl group, the apparent one-bond correlation arises from the $^3J_{C-H}$ interaction of the protons of one of the methyl groups with the chemically equivalent carbon which is three bonds away.

10. The *tert*-butyl protons (H_8) also correlate to the quaternary carbon C_7 at 35.1 ppm, and the quaternary aromatic carbon at 156.8 ppm. The resonance at 156.8 ppm is therefore assigned to C_4.

11. The proton resonance at 7.90 ppm also correlates to C_4. This proton resonance and its associated carbon resonance at 128.3 ppm are therefore identified as H_2 and C_2. H_3 must be the proton resonance at 7.48 ppm and C_3 the carbon resonance at 125.5 ppm.

12. The remaining ^{13}C resonance at 134.6 ppm is assigned to C_1.

^{1}H–^{13}C HMBC spectrum of 4'-*tert*-butylacetophenone (CDCl$_3$, 300 MHz)

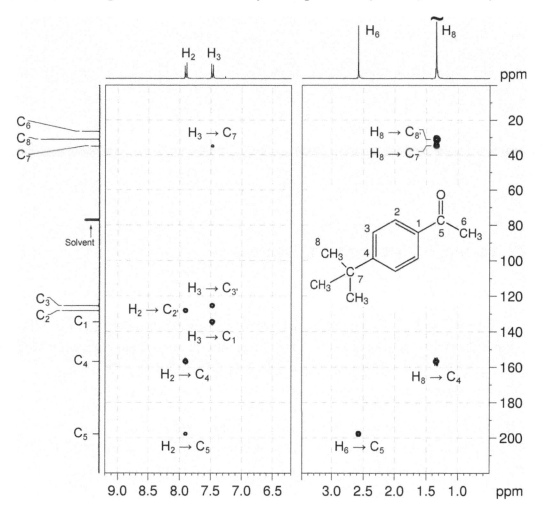

13. In the HMBC spectrum for this compound, there are strong correlations between H$_2$ and C$_{2'}$ and also between H$_3$ and C$_{3'}$. While these appear to be one-bond correlations, in *para*-disubstituted benzenes (and indeed in any substituted aromatic ring in which a mirror plane bisects two of the substituents), these apparent one-bond correlations arise from the $^3J_{C–H}$ interaction of a proton with the carbon which is *meta* to it.

Question:

Identify the following compound.

Molecular Formula: $C_{12}H_{16}O$

IR Spectrum: 1685 cm^{-1}

Solution:

2,2,4'-Trimethylpropiophenone

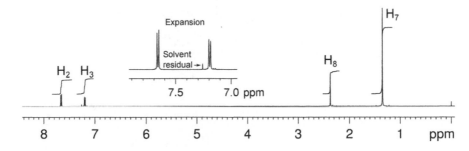

1. The molecular formula is $C_{12}H_{16}O$. Calculate the degree of unsaturation from the molecular formula: ignore the O atom to give an effective molecular formula of $C_{12}H_{16}$ (C_nH_m) which gives the degree of unsaturation as $(n - m/2 + 1) = 12 - 8 + 1 = 5$. The compound contains a combined total of five rings and/or π bonds.

2. The IR and 1D NMR spectra establish that the compound is aromatic and contains a ketone functional group (quaternary ^{13}C resonance at 208.3 ppm). This accounts for all of the degrees of unsaturation, so the compound contains no additional rings or multiple bonds.

3. There are two aromatic proton resonances at 7.66 and 7.19 ppm in the ^1H NMR spectrum. On the basis of the coupling pattern, the ring is *para*-disubstituted, but one cannot readily distinguish which resonance belongs to H_2 and which to H_3.

4. The ^1H NMR spectrum has one three-proton resonance at 2.38 ppm (H_8), and a nine-proton resonance at 1.35 ppm (H_7). The chemical shift of the resonance at 2.38 ppm is consistent with a methyl group either bound directly to the aromatic ring or to the ketone, while the resonance at 1.35 ppm is due to a *tert*-butyl group.

^1H NMR spectrum of 2,2,4'-trimethylpropiophenone (CDCl$_3$, 500 MHz)

$^{13}C\{^1H\}$ NMR spectrum of 2,2,4'-trimethylpropiophenone (CDCl$_3$, 125 MHz)

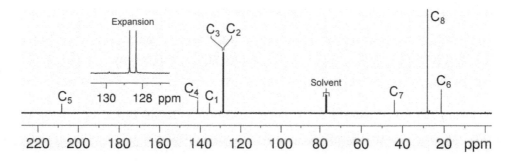

5. There are two possible isomers:

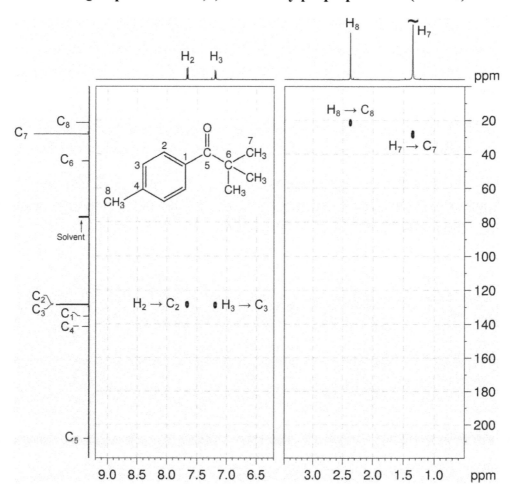

6. The ^1H–^{13}C me-HSQC identifies the protonated carbon resonances: C_7 at 28.2 ppm, C_8 at 21.4 ppm and the protonated aromatic carbons at 128.3 and 128.7 ppm.

^1H–^{13}C me-HSQC spectrum of 2,2,4'-trimethylpropiophenone (CDCl$_3$, 500 MHz)

7. The ^{1}H–^{13}C HMBC spectrum has a correlation between H$_7$ (the *tert*-butyl group) and the carbonyl carbon C$_5$. This identifies **Isomer B** as the correct structure.

8. The HMBC spectrum can be used to identify and assign the remaining carbon and proton resonances. **Remember** that, in aromatic systems, the three-bond coupling $^3J_{C-H}$ is typically the larger long-range coupling and gives rise to the strongest cross-peaks.

9. Note that the *tert*-butyl protons (H$_7$) correlate to the *tert*-butyl methyl carbon resonance at 28.2 ppm (C$_7'$) in the HMBC spectrum. While this appears to be a one-bond correlation, in *tert*-butyl groups, isopropyl groups or compounds with a *gem*-dimethyl group, the apparent one-bond correlation arises from the $^3J_{C-H}$ interaction of the protons of one of the methyl groups with the chemically equivalent carbon which is three bonds away.

10. The *tert*-butyl protons (H$_7$) also correlate to the aliphatic quaternary carbon C$_6$ at 44.0 ppm in the HMBC spectrum.

11. In the HMBC spectrum, the aromatic proton resonance at 7.19 ppm correlates to the methyl carbon resonance at 21.4 ppm (C$_8$). The resonance at 7.19 ppm can thus be assigned to the aromatic proton adjacent to the methyl group (H$_3$).

12. The remaining aromatic proton signal at 7.66 ppm must be due to H$_2$. Both C$_2$ and C$_3$ can be assigned to the signals at 128.3 and 128.7 ppm, respectively using the me-HSQC spectrum.

13. In the HMBC spectrum, H$_3$ also correlates to the quaternary aromatic carbon signal at 135.4 ppm and this signal can be assigned to C$_1$ as it is three bonds away from H$_3$.

14. In the HMBC spectrum, H$_2$ correlates to the quaternary aromatic carbon signal at 141.5 ppm which can now be assigned to C$_4$. H$_2$ also correlates to the carbonyl carbon C$_5$ confirming that Isomer B is the correct structure.

¹H–¹³C HMBC spectrum of 2,2,4'-trimethylpropiophenone (CDCl₃, 500 MHz)

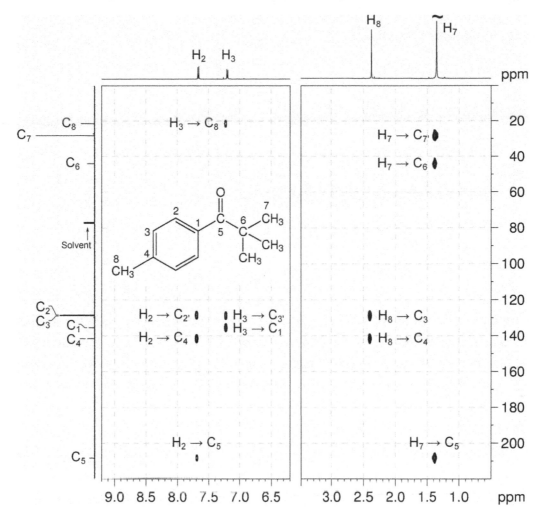

15. In the HMBC spectrum for this compound, there are strong correlations between H_2 and $C_{2'}$ and also between H_3 and $C_{3'}$. While these appear to be one-bond correlations, in *para*-disubstituted benzenes (and in any substituted aromatic ring in which a mirror plane bisects two of the substituents), these apparent one-bond correlations arise from the $^3J_{C-H}$ interaction of a proton with the carbon which is *meta* to it.

Problem 24

Question:

Identify the following compound.

Molecular Formula: $C_{10}H_{12}O$

IR: 3320 (br), 1600 (w), 1492 (m), 1445 (m) cm^{-1}

Solution:

trans-2-Methyl-3-phenyl-2-propen-1-ol

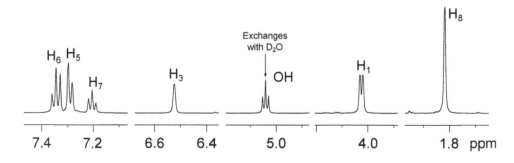

1. The molecular formula is $C_{10}H_{12}O$. Calculate the degree of unsaturation from the molecular formula: ignore the O atom to give an effective molecular formula of $C_{10}H_{12}$ (C_nH_m) which gives the degree of unsaturation as $(n - m/2 + 1) = 10 - 6 + 1 = 5$. The compound contains a combined total of five rings and/or π bonds.

2. In the ^1H NMR spectrum, the signal at 5.04 ppm exchanges with D_2O so this must be due to an –OH group. The IR stretch at 3320 cm^{-1} and the fact that this signal has no correlation to a carbon signal in the me-HSQC spectrum verifies the presence of an OH group.

3. In the ^1H NMR spectrum, the integration and coupling pattern of the aromatic protons indicate the presence of a mono-substituted benzene ring. The signal at 6.52 ppm is in the alkene region, whilst the signals at 4.02 and 1.82 ppm indicate –OCH_2– and –CH_3 fragments, respectively.

4. The aromatic ring and alkene functional group accounts for all of the degrees of unsaturation, so the compound contains no additional rings or multiple bonds.

^1H NMR spectrum of *trans*-2-methyl-3-phenyl-2-propen-1-ol (DMSO-d_6, 500 MHz)

$^{13}C\{^1H\}$ **NMR spectrum of** *trans*-**2-methyl-3-phenyl-2-propen-1-ol (DMSO-***d_6***, 125 MHz)**

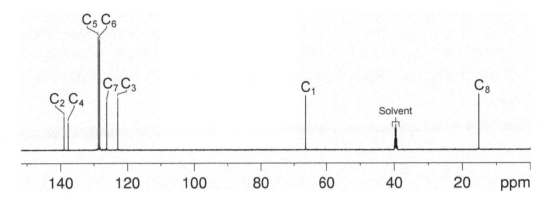

5. The COSY spectrum shows that the signals for the OH and CH_2 groups are coupled thus they must be directly bound.

^1H–^1H **COSY spectrum of** *trans*-**2-methyl-3-phenyl-2-propen-1-ol (DMSO-***d_6***, 500 MHz)**

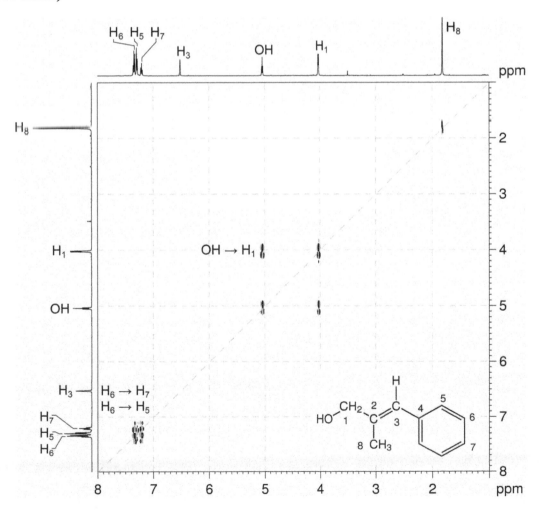

6. The HSQC spectrum easily identifies the protonated carbons (aromatic carbons C_5, C_6 and C_7 at 128.6, 128.2 and 126.1, respectively; alkene carbon C_3 at 122.8, OCH_2 at 66.7 and CH_3 at 15.2 ppm).

^1H–^{13}C me-HSQC spectrum of *trans*-2-methyl-3-phenyl-2-propen-1-ol (DMSO-d_6, 500 MHz)

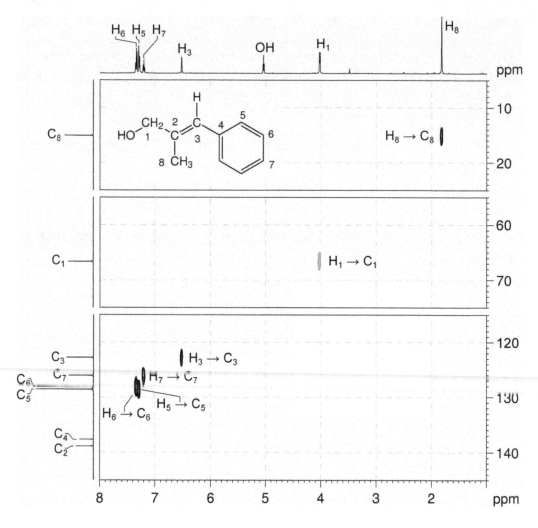

7. In the HMBC spectrum, there are correlations from the =CH, OH, CH_2 and CH_3 groups to a quaternary carbon C_2 which has a chemical shift in the aromatic / alkene region (138.9 ppm). Joining C_2 to the =CH group affords a $HO–CH_2–C(CH_3)=CH$ fragment.

8. The HMBC spectrum can be used to assign the *ipso* carbon C_4 at 137.7 ppm ($H_6–C_4$ correlation).

9. Correlations from alkene proton H_3 to *ipso* carbon C_4 and aromatic carbon C_5 indicate that the phenyl group is attached to the =CH group.

$^1H–^{13}C$ HMBC spectrum of *trans*-2-methyl-3-phenyl-2-propen-1-ol (DMSO-d_6, 500 MHz)

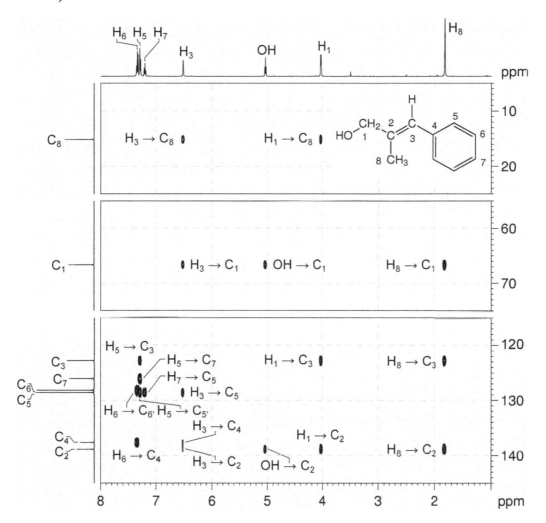

10. In the HMBC spectrum for this compound, there are strong correlations between H_5 and $C_{5'}$ and also between H_6 and $C_{6'}$. While these appear to be one-bond correlations, in any substituted aromatic ring in which a mirror plane bisects two of the substituents, these apparent one-bond correlations arise from the $^3J_{C–H}$ interaction of a proton with the carbon which is *meta* to it.

11. The configuration about the double bond can be deduced from the $^1H-^1H$ NOESY spectrum. The H_3-H_1 and H_3-OH correlations indicate that the alkene proton is on the same side of the double bond as the CH_2OH group. Similarly the H_8-H_5 correlation indicates that the methyl and phenyl groups are on the same side of the double bond. The absence of correlations between the methyl group (H_8) and the alkene proton (H_3) confirms that these groups are *trans* to each other across the double bond.

$^1H-^1H$ NOESY spectrum *trans*-2-methyl-3-phenyl-2-propen-1-ol (DMSO-d_6, 500 MHz)

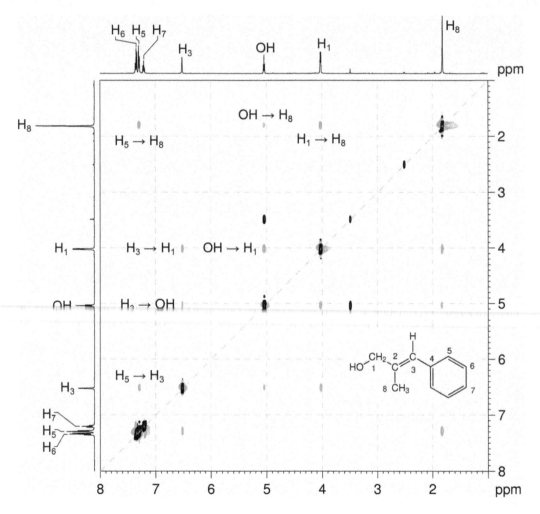

Question:

Identify the following compound.

Molecular Formula: $C_{10}H_{12}O_3$

IR: 1720 cm^{-1}

Solution:

Methyl 4-ethoxybenzoate

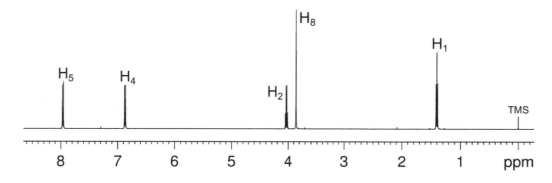

1. The molecular formula is $C_{10}H_{12}O_3$. Calculate the degree of unsaturation from the molecular formula: ignore the O atoms to give an effective molecular formula of $C_{10}H_{12}$ (C_nH_m) which gives the degree of unsaturation as $(n - m/2 + 1) = 10 - 6 + 1 = 5$. The compound contains a combined total of five rings and/or π bonds.

2. The aromatic signals in the ^{1}H NMR spectrum indicate the presence of a benzene ring which accounts for four degrees of unsaturation. The remaining degree of unsaturation must be due to an ester group as also confirmed by IR and quaternary ^{13}C resonance at 166.8 or 162.8 ppm.

3. In the ^{1}H NMR spectrum, there are two aromatic signals at 7.96 and 6.87 ppm (H_5 and H_4) with coupling patterns typical of a *para*-disubstituted benzene ring. There are also signals for methoxy (3.86 ppm, H_8) and ethoxy (4.03 and 1.40 ppm, H_2 and H_1) groups.

^{1}H NMR spectrum of methyl 4-ethoxybenzoate (CDCl$_3$, 500 MHz)

$^{13}C\{^1H\}$ NMR spectrum of methyl 4-ethoxybenzoate (CDCl$_3$, 125 MHz)

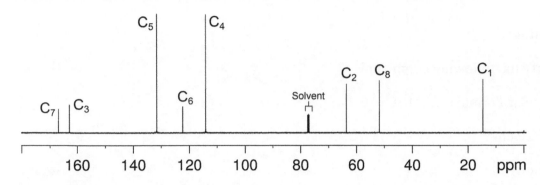

4. The COSY spectrum shows the expected correlations between the two aromatic signals H$_5$ and H$_4$ as well as the correlations between the OCH$_2$ (H$_2$) and CH$_3$ (H$_1$) resonances.

^1H–^1H COSY spectrum of methyl 4-ethoxybenzoate (CDCl$_3$, 500 MHz)

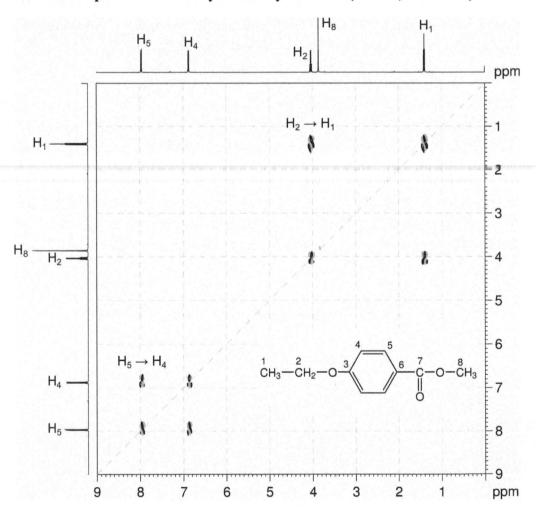

5. The me-HSQC spectrum easily identifies the protonated carbons. The protonated aromatic carbons are at 131.6 and 114.1 ppm. The OCH_2 carbon is at 63.7 ppm, the OCH_3 carbon is at 51.8 ppm and the CH_3 carbon is at 14.9 ppm.

^1H–^{13}C me-HSQC spectrum of methyl 4-ethoxybenzoate (CDCl$_3$, 500 MHz)

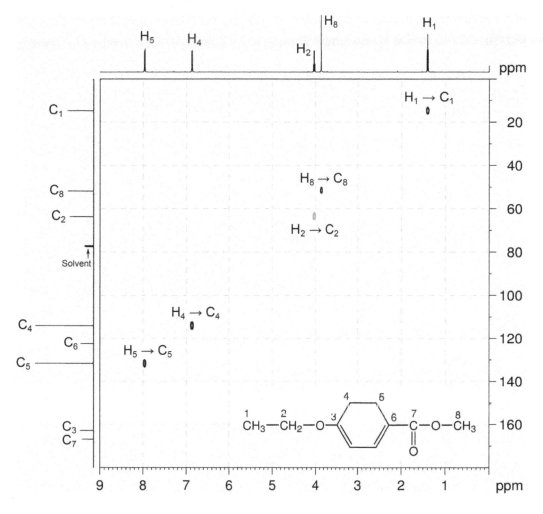

6. There are two possible isomers – one with the ethoxy group as a substituent on the benzene ring and the methoxy group as part of the ester (isomer A) and the other with the methoxy group as the substituent on the benzene ring and the ethoxy group as part of the ester (isomer B).

7. In the HMBC spectrum, the correlations between aromatic protons H_4 and H_5 with the ^{13}C resonance at 162.8 ppm indicate that this ^{13}C resonance is due to the quaternary aromatic carbon C_3 which is shifted downfield because it is bonded to an oxygen atom. Thus, the carbonyl carbon C_7 must be the signal at 166.8 ppm.

8. In the HMBC spectrum, the strong correlation between the methoxy protons H_8 and the carbonyl carbon C_7 identifies the correct structure as **Isomer A**. The correlation between ethoxy proton H_2 with quaternary aromatic carbon C_3 confirms that the ethoxy group is bound directly to the benzene ring.

$^1H–^{13}C$ HMBC spectrum of methyl 4-ethoxybenzoate – expansion A (CDCl$_3$, 500 MHz)

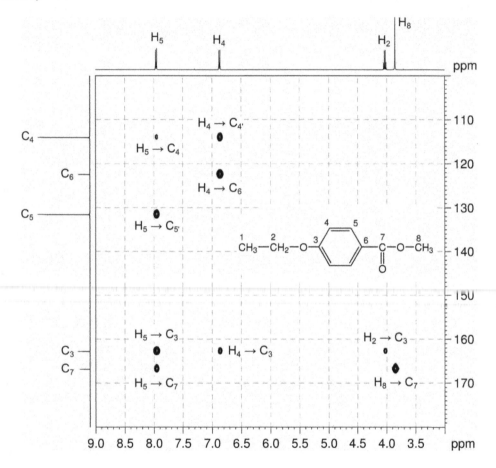

9. In the HMBC spectrum for this compound, there are strong correlations between H_4 and $C_{4'}$ and also between H_5 and $C_{5'}$. While these appear to be one-bond correlations, in *para*-disubstituted benzenes (and in any substituted aromatic ring in which a mirror plane bisects two of the substituents), these apparent one-bond correlations arise from the $^3J_{C–H}$ interaction of a proton with the carbon which is *meta* to it.

Problem 26

Question:

Identify the following compound.

Molecular Formula: $C_{11}H_{14}O_2$

IR: 1739 cm^{-1}

Solution:

Methyl 3-(*p*-tolyl)propionate

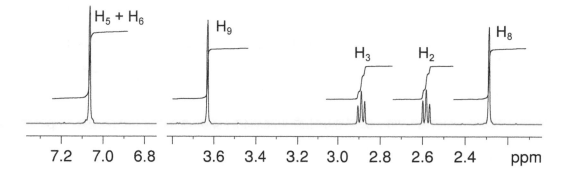

1. The molecular formula is $C_{11}H_{14}O_2$. Calculate the degree of unsaturation from the molecular formula: ignore the O atoms to give an effective molecular formula of $C_{11}H_{14}$ (C_nH_m) which gives the degree of unsaturation as $(n - m/2 + 1) = 11 - 7 + 1 = 5$. The compound contains a combined total of five rings and/or π bonds.

2. IR and 1D NMR spectra establish that the compound is aromatic and contains an ester functional group (quaternary ^{13}C resonance at 173.3 ppm). This accounts for all of the degrees of unsaturation, so the compound contains no additional rings or multiple bonds.

3. In the ^1H NMR spectrum, the aromatic signal at 7.06 ppm integrates for four protons ($H_5 + H_6$) and the lack of any obvious coupling pattern indicates that all the aromatic protons are in a similar environment – probably a *para*-disubstituted benzene ring where the shift of all protons is similar. There are also signals for methoxy –OCH_3 (3.63 ppm, H_9), –CH_2CH_2– (2.89 and 2.58 ppm, H_3 and H_2) and methyl –CH_3 (2.29 ppm, H_8) groups.

^1H NMR spectrum of methyl 3-(*p*-tolyl)propionate (CDCl₃, 500 MHz)

$^{13}C\{^1H\}$ NMR spectrum of methyl 3-(*p*-tolyl)propionate (CDCl₃, 125 MHz)

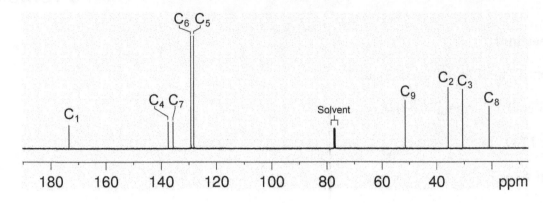

4. The COSY spectrum shows the expected correlations between the two –CH₂– signals H₃ and H₂.

^1H–^1H COSY spectrum of methyl 3-(*p*-tolyl)propionate (CDCl₃, 500 MHz)

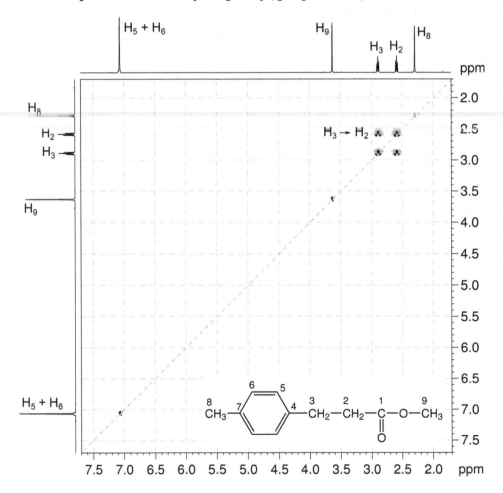

5. The me-HSQC spectrum easily identifies the protonated carbons. Note that the ^1H signal for $H_5 + H_6$ correlates to two C signals at 129.2 and 128.2 ppm for C_6 and C_5. The OCH_3 carbon is at 51.5 ppm, the two CH_2 carbons are at 35.9 and 30.6 ppm, and the CH_3 carbon is at 21.0 ppm.

^1H–^{13}C me-HSQC spectrum of methyl 3-(*p*-tolyl)propionate (CDCl3, 500 MHz)

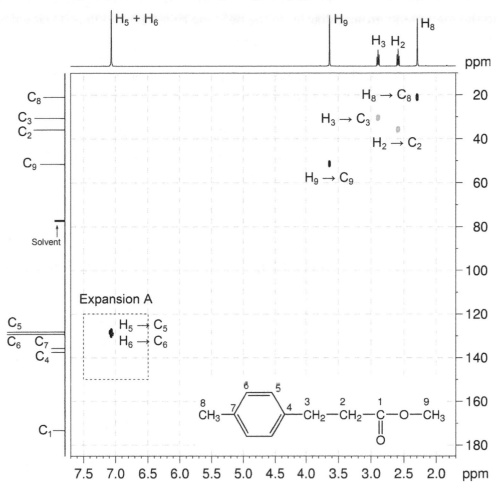

6. In the HMBC spectrum, the correlation between methoxy protons H_9 and carbonyl carbon C_1 indicates the presence of a methoxy ester. Correlations between methylene protons H_2 and H_3 with carbonyl carbon C_1 indicate that the $-CH_2CH_2-$ group is also bound to the carbonyl group.

7. In the HMBC spectrum, the correlation from methylene proton signal at 2.89 ppm to protonated aromatic carbon at 128.2 ppm signals indicate that the $-CH_2CH_2-$ group is bound to the benzene ring. The correlation also identifies the ^1H signal at 2.89 ppm as due to the methylene group adjacent to the benzene ring (H_3) and the ^{13}C signal at 128.2 ppm as to the protonated aromatic carbon *ortho* to the methylene substituent (C_5).

8. In the HMBC spectrum, H_3 also correlates to quaternary aromatic carbon signal at 137.5 ppm which can be assigned to C_4.

9. The methyl group must be bound to the benzene ring as shown by correlations from the methyl proton signals at 2.29 ppm to protonated aromatic carbon signal at 129.2 ppm (which can be assigned to C_6) and quaternary aromatic carbon signal at 135.7 ppm (which can be assigned to C_7).

^1H–^{13}C HMBC spectrum of methyl 3-(*p*-tolyl)propionate (CDCl$_3$, 500 MHz)

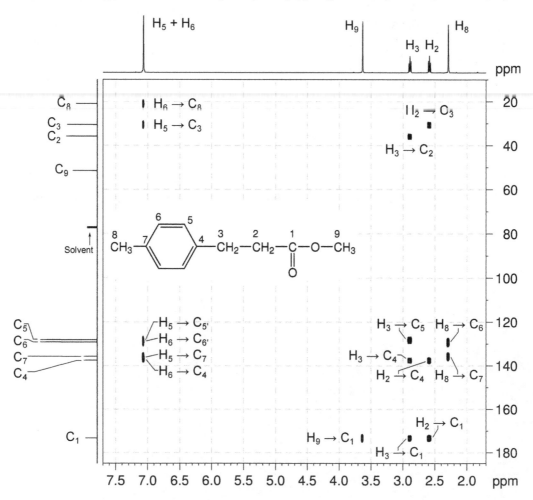

Problem 27

Question:

Identify the following compound.

Molecular Formula: $C_{11}H_{14}O_2$

IR: 1715 cm^{-1}

Draw a labeled structure, and use the me-HSQC and INADEQUATE spectra to assign each ^1H and ^{13}C resonance to the corresponding nucleus in the structure.

Solution:

4-(4'-Methoxyphenyl)-2-butanone $CH_3-O-\overset{8}{\underset{}{}}\!\!\!\overset{7\!\!-\!\!6}{\bigcirc}\!\!\overset{5}{}-CH_2-CH_2-\underset{\underset{O}{\|}}{C}-CH_3$

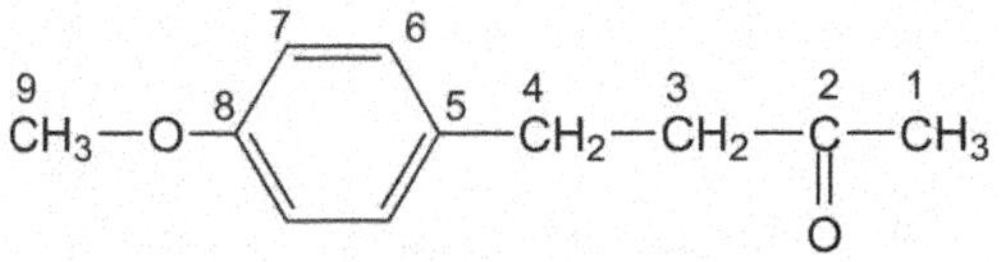

Proton	Chemical Shift (ppm)	Carbon	Chemical Shift (ppm)
H$_1$	2.09	C$_1$	30.0
		C$_2$	208.0
H$_3$	2.69	C$_3$	45.9
H$_4$	2.81	C$_4$	28.9
		C$_5$	133.0
H$_6$	7.08	C$_6$	129.2
H$_7$	6.80	C$_7$	113.9
		C$_8$	158.0
H$_9$	3.74	C$_9$	55.2

1. The molecular formula is $C_{11}H_{14}O_2$. Calculate the degree of unsaturation from the molecular formula: ignore the O atoms to give an effective molecular formula of $C_{11}H_{14}$ (C_nH_m) which gives the degree of unsaturation as $(n - m/2 + 1) = 11 - 7 + 1 = 5$. The compound contains a combined total of five rings and/or π bonds.

2. IR and 1D NMR spectra establish that the compound is aromatic and contains a ketone functional group (quaternary ^{13}C resonance at 208.0 ppm). This accounts for all of the degrees of unsaturation and there are no additional rings or multiple bonds.

3. In the ^1H NMR spectrum, there are two aromatic signals at 7.08 and 6.80 ppm (H$_6$ and H$_7$) with coupling patterns typical of a *para*-disubstituted benzene ring. There

are also signals for methoxy (3.74 ppm, H_9), $-CH_2CH_2-$ (2.81 and 2.69 ppm, H_4 and H_3) and methyl (2.09 ppm, H_1) groups.

^{1}H NMR spectrum of 4-(4'-methoxyphenyl)-2-butanone (CDCl$_3$, 600 MHz)

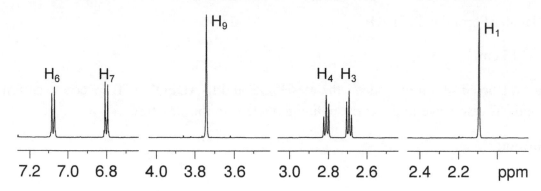

^{13}C{^{1}H} NMR spectrum of 4-(4'-methoxyphenyl)-2-butanone (CDCl$_3$, 150 MHz)

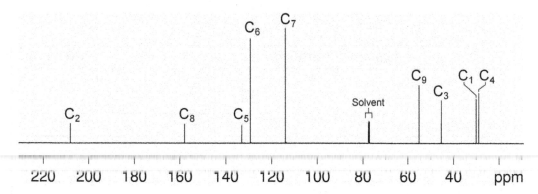

4. The me-HSQC spectrum easily identifies the signals for the protonated carbons – the protonated aromatic carbons are at 129.2 and 113.9 ppm (C_6 and C_7), the methoxy carbon is at 55.2 ppm (C_9), the two methylene carbons are at 45.4 and 28.9 ppm (C_3 and C_4), and the methyl carbon is at 30.1 ppm (C_4). The signals at 158.0 and 133.0 ppm belong to quaternary aromatic carbons (C_5 and C_8) as these signals have no correlations in the me-HSQC spectrum.

^{1}H–^{13}C me-HSQC spectrum of 4-(4'-methoxyphenyl)-2-butanone (CDCl$_3$, 600 MHz)

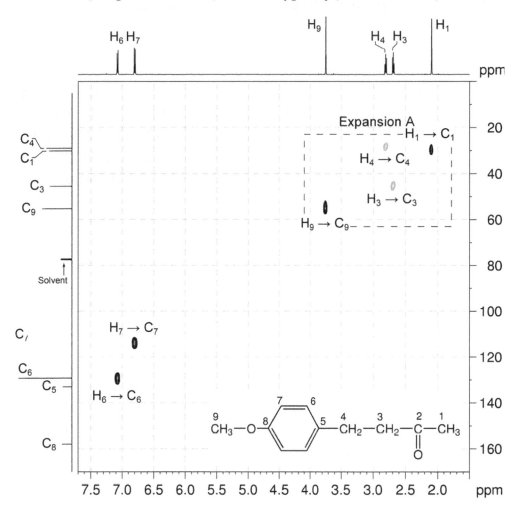

5. The INADEQUATE spectrum shows direct $^{13}C-^{13}C$ connectivity.

6. Beginning, for example, with the ketone carbon resonance at 208.0 ppm (C_2), correlations to the methyl carbon signal at 30.0 ppm (C_1) and methylene carbon signal at 45.9 ppm (C_3) affords the following fragment:

$$-CH_2-\underset{\underset{O}{\|}}{C}-CH_3$$

7. The methylene carbon signal at 45.9 ppm (C_3) has a correlation with the methylene carbon signal at 28.9 ppm (C_4) which extends the fragment to:

$$-CH_2-CH_2-\underset{\underset{O}{\|}}{C}-CH_3$$

8. The methylene carbon signal at 28.9 ppm (C_4) has a correlation with the quaternary aromatic carbon signal at 133.0 ppm (C_5). The signal at 133.0 ppm (C_5) has a correlation with the protonated aromatic carbon signal at 129.2 ppm (C_6). The signal at 129.2 ppm (C_6) has a correlation with the protonated aromatic carbon signal at 113.9 ppm (C_7) which in turn has a correlation with the quaternary aromatic carbon signal at 158.0 ppm (C_8). We now have the following fragment:

9. The remaining methoxy group must be bound to the benzene ring at the position *para* to the other substituent. There are no correlations between the methoxy carbon signal at 55.2 ppm (C_9) and the quaternary aromatic signal at 158.0 ppm (C_8) due to the presence of an oxygen atom between then.

INADEQUATE spectrum of 4-(4'-methoxyphenyl)-2-butanone (CDCl₃, 150 MHz)

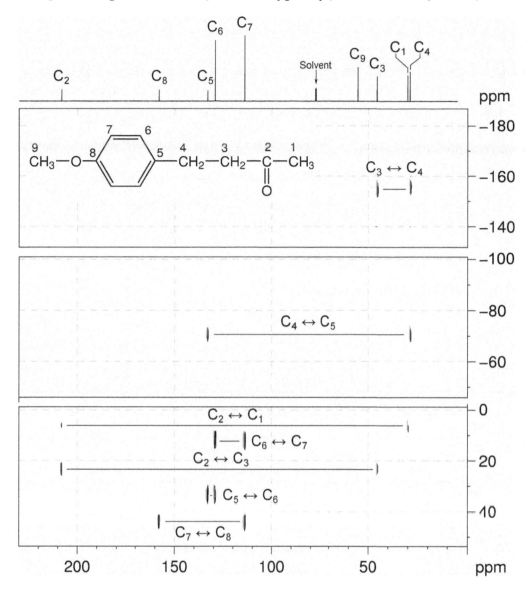

Problem 28

Question:

The ^{1}H and ^{13}C$\{^1$H$\}$ NMR spectra of ethyl 6-bromohexanoate ($C_8H_{15}BrO_2$) recorded in CDCl$_3$ solution at 298 K and 500 MHz are given below. The ^{1}H NMR spectrum has signals at δ 1.25, 1.48, 1.65, 1.87, 2.31, 3.41 and 4.12 ppm. The ^{13}C$\{^1$H$\}$ NMR spectrum has signals at δ 14.3, 24.0, 27.6, 32.4, 33.5, 34.0, 60.2 and 173.3 ppm. The 2D ^{1}H–^{1}H COSY and the multiplicity-edited ^{1}H–^{13}C HSQC spectra are given on the facing page. From the COSY spectrum, assign the proton spectrum, then use this information to assign the ^{13}C$\{^1$H$\}$ spectrum.

Solution:

Ethyl 6-bromohexanoate

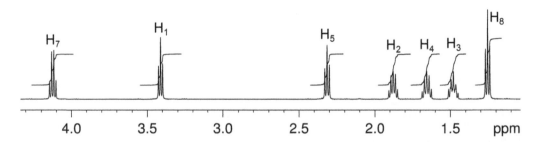

Proton	Chemical Shift (ppm)	Carbon	Chemical Shift (ppm)
H$_1$	3.41	C$_1$	33.5
H$_2$	1.87	C$_2$	32.4
H$_3$	1.48	C$_3$	27.6
H$_4$	1.65	C$_4$	24.0
H$_5$	2.31	C$_5$	34.0
		C$_6$	173.3
H$_7$	4.12	C$_7$	60.2
H$_8$	1.25	C$_8$	14.3

^{1}H NMR spectrum of ethyl 6-bromohexanoate (CDCl$_3$, 500 MHz)

$^{13}C\{^{1}H\}$ NMR spectrum of ethyl 6-bromohexanoate (CDCl$_3$, 125 MHz)

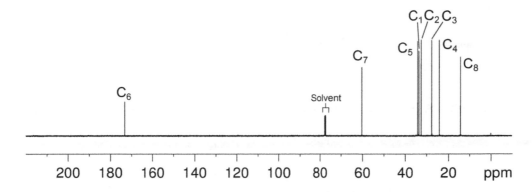

1. Two separate spin systems can be observed in the COSY spectrum – one for the ethoxy group and another for the butyl chain. The ethoxy (–OCH$_2$CH$_3$) group can be identified and H$_7$ and H$_8$ assigned to the resonances at 4.12 and 1.25 ppm, respectively.

2. H$_1$ and H$_5$ can be assigned to the resonances at 3.41 and 2.31 ppm, respectively, on the basis of chemical shift and coupling pattern.

3. In the COSY spectrum, the H$_1$–H$_2$, H$_2$–H$_3$, H$_3$–H$_4$ and H$_4$–H$_5$ correlations allow H$_2$, H$_3$ and H$_4$ to be assigned to the signals at 1.87, 1.48 and 1.65 ppm, respectively.

^{1}H–^{1}H COSY spectrum of ethyl 6-bromohexanoate (CDCl$_3$, 500 MHz)

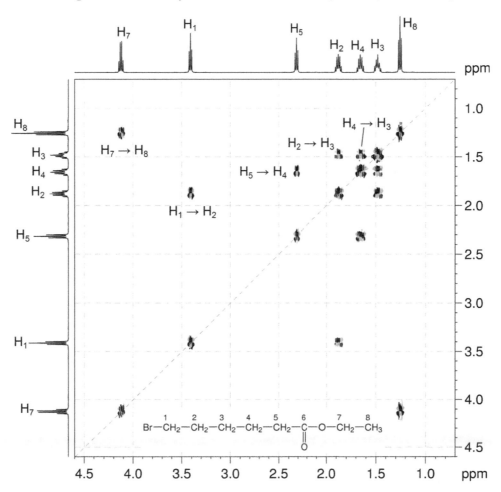

4. The carbonyl carbon C_6 can be assigned easily to the ^{13}C resonance at 173.3 ppm on the basis of chemical shift.

5. The remaining C resonances can be easily assigned by the me-HSQC spectrum.

^1H–^{13}C me-HSQC spectrum of ethyl 6-bromohexanoate (CDCl$_3$, 500 MHz)

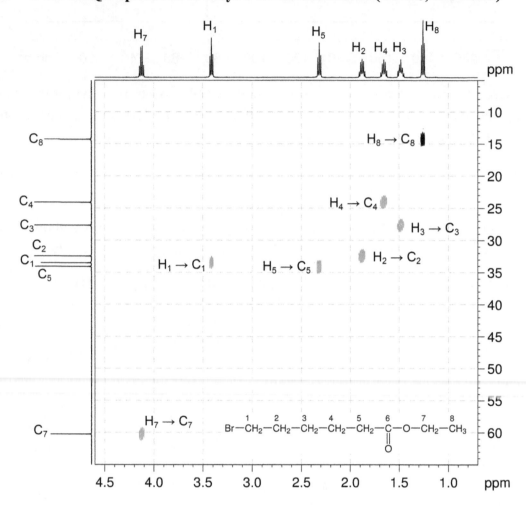

Question:

The ^{1}H and ^{13}C{^{1}H} NMR spectra of piperonal ($C_8H_6O_3$) recorded in CDCl$_3$ solution at 298 K and 600 MHz are given below.

The ^{1}H NMR spectrum has signals at δ 6.07 (s, 2H, H$_7$), 6.92 (d, $^3J_{H-H}$ = 7.9 Hz, 1H, H$_1$), 7.31 (d, $^4J_{H-H}$ = 1.5 Hz, 1H, H$_4$), 7.40 (dd, $^3J_{H-H}$ = 7.9 Hz, $^4J_{H-H}$ = 1.5 Hz, 1H, H$_6$) and 9.80 (s, 1H, H$_8$) ppm.

The ^{13}C{^{1}H} NMR spectrum has signals at δ 102.1 (C$_7$), 106.9 (C$_4$), 108.3 (C$_1$), 128.6 (C$_6$), 131.9 (C$_5$), 148.7 (C$_3$), 153.1 (C$_2$) and 190.2 (C$_8$) ppm.

Use this information to produce schematic diagrams of the HSQC and HMBC spectra, showing where all of the cross-peaks and diagonal peaks would be.

Solution:

Piperonal

1. The assignments for the ^{1}H and ^{13}C{^{1}H} NMR spectra are given.

^{1}H NMR spectrum of piperonal (CDCl$_3$, 600 MHz)

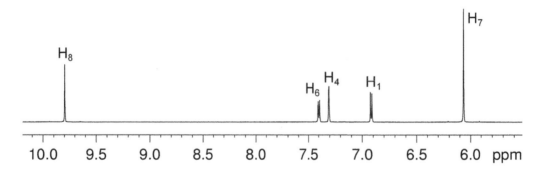

^{13}C{^{1}H} NMR spectrum of piperonal (CDCl$_3$, 150 MHz)

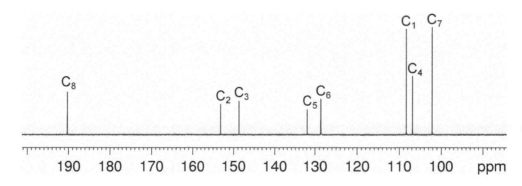

135

2. The ^1H–^{13}C me-HSQC spectrum shows direct (one-bond) correlations between proton and carbon nuclei, so there will be cross-peaks between H_1 and C_1, H_4 and C_4, H_6 and C_6, H_7 and C_7 (shown here in red, as there are two protons directly connected to C_7) and between H_8 and C_8.

Predicted ^1H–^{13}C me-HSQC spectrum of piperonal (CDCl3, 600 MHz)

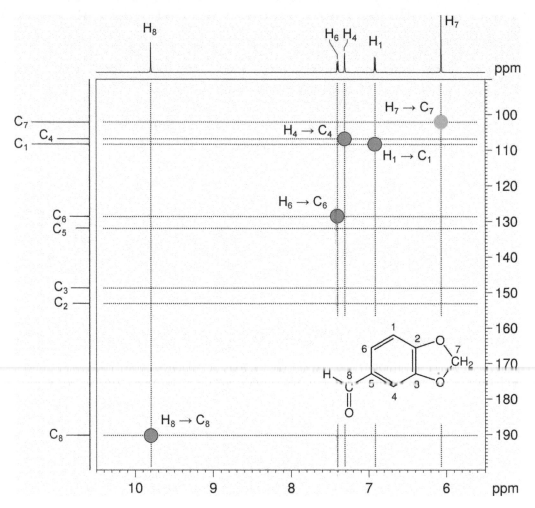

3. In HMBC spectra, remember that, for aromatic systems, the three-bond coupling $^3J_{C-H}$ is typically the larger long-range coupling and gives rise to the strongest cross-peaks. Benzylic protons correlate to the *ipso* carbon (two-bond correlation) and the *ortho* carbons (three-bond correlations).

Predicted ^{1}H–^{13}C HMBC spectrum of piperonal (CDCl$_3$, 600 MHz)

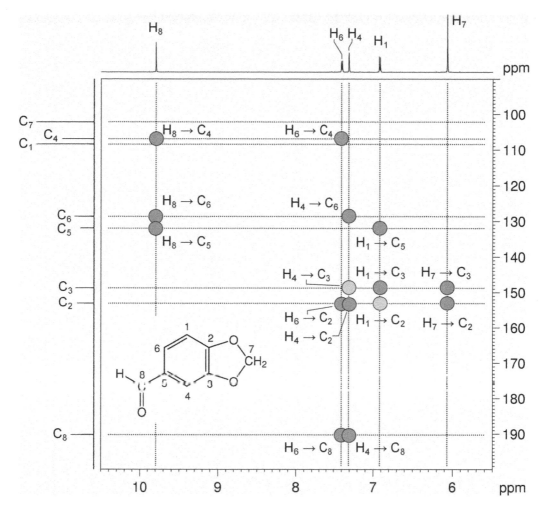

4. H_1 will correlate to the carbon nuclei three bonds removed (*i.e. meta*): C_3 and C_5. Oxygen substituents on aromatic rings increase the magnitude of two-bond C–H coupling to the oxygen-substituted carbon atom from ≤ 1 Hz to ≥ 2.5 Hz. It is therefore often possible to observe two-bond correlations to the phenolic *ipso* carbon, and indeed H_1 does show such a correlation (shown in light grey).

5. H_4 will correlate to the *meta* carbon nuclei: C_2 and C_6. There will also be a three-bond correlation to the aldehydic carbon atom (C_8). In addition, for the reasons outlined above, there is a two-bond correlation to C_3 (shown in light grey).

6. H_6 will correlate to the *meta* carbon nuclei: C_2 and C_4. There will also be a three-bond correlation to the aldehydic carbon atom (C_8).

7. H_7 will, in principle, show both two- and three-bond correlations; however there are no carbon nuclei which are two bonds removed from H_7, so we expect only three-bond correlations to C_2 and C_3.

8. H_8 will also show both two- and three-bond correlations. There will be a two-bond correlation to C_5, and three-bond correlations to both C_4 and C_6.

Question:

Identify the following compound.

Molecular Formula: $C_{13}H_{16}O_2$

IR: 1720 cm^{-1}

Solution:

cis-3-Hexenyl benzoate

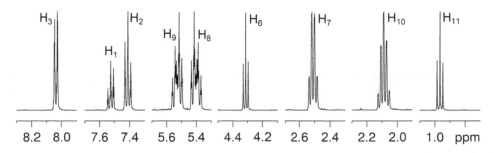

1. The molecular formula is $C_{13}H_{16}O_2$. Calculate the degree of unsaturation from the molecular formula: ignore the O atoms to give an effective molecular formula of $C_{13}H_{16}$ (C_nH_m) which gives the degree of unsaturation as $(n - m/2 + 1) = 13 - 8 + 1 = 6$. The compound contains a combined total of six rings and/or π bonds.

2. 1D NMR spectra establish the presence of a mono-substituted benzene ring (^1H signals at 8.04, 7.53 and 7.41 ppm for H_3, H_1 and H_2, respectively), two alkene protons (^1H signals at 5.53 and 5.40 ppm for H_9 and H_8, respectively) and an ester group (^{13}C signal at 166.6 ppm for C_5).

3. The aromatic, alkene and ester functional groups account for all of the degrees of unsaturation, so the compound contains no additional rings or multiple bonds.

4. In the ^1H NMR spectrum, there are also signals for one methoxy group at 4.31 ppm (H_6), two methylene groups at 2.51 and 2.09 ppm (H_7 and H_{10}) and a methyl group at 0.97 ppm (H_{11}).

^1H NMR spectrum of *cis*-3-hexenyl benzoate (CDCl$_3$, 400 MHz)

$^{13}C\{^1H\}$ NMR spectrum of *cis*-3-hexenyl benzoate (CDCl$_3$, 100 MHz)

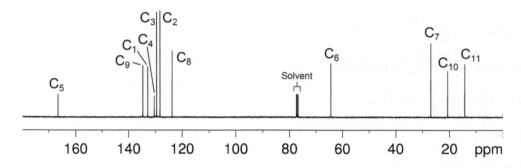

5. In the COSY spectrum, there are two spin systems – one of which corresponds to the aromatic ring (H$_2$–H$_3$ and H$_1$–H$_2$). The other spin system, starting with the methyl group (H$_{11}$) which correlates to a –CH$_2$– group (H$_{10}$) which in turn correlates to a –CH= group (H$_9$). The correlations continue with –CH= group (H$_8$) then –CH$_2$– group (H$_7$) and –OCH$_2$ group (H$_6$). These correlations afford the CH$_3$CH$_2$CH=CHCH$_2$CH$_2$O– fragment.

^1H–^1H COSY spectrum of *cis*-3-hexenyl benzoate (CDCl$_3$, 400 MHz)

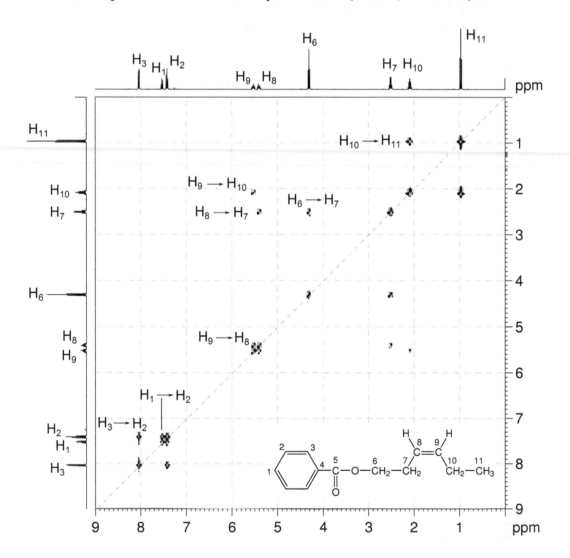

6. The protonated carbons C_6, C_7, C_{10} and C_{11} are easily identified by the me-HSQC spectrum as signals at 64.5, 26.9, 20.7 and 14.3 ppm, respectively.

^{1}H–^{13}C me-HSQC spectrum of *cis*-3-hexenyl benzoate (CDCl$_3$, 400 MHz)

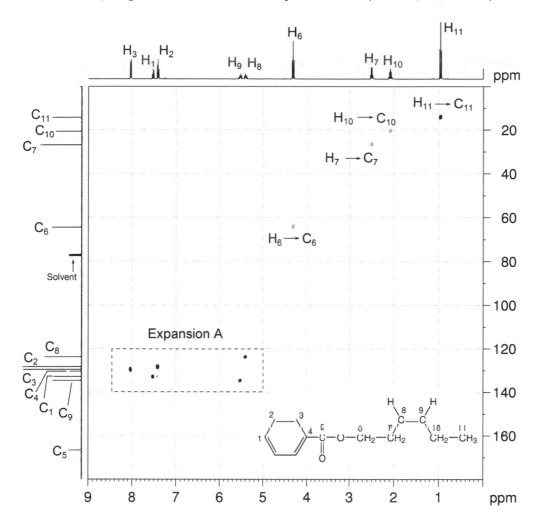

7. The protonated carbons C_1, C_2, C_3, C_8 and C_9 are easily identified by the me-HSQC spectrum (expansion A) as signals at 132.9, 128.3, 129.6, 123.8 and 134.6 ppm, respectively.

8. The remaining ^{13}C signal at 130.5 ppm does not have a correlation in the me-HSQC spectrum and must be due to the quaternary aromatic carbon C_4.

^1H–^{13}C me-HSQC spectrum of *cis*-3-hexenyl benzoate – expansion A

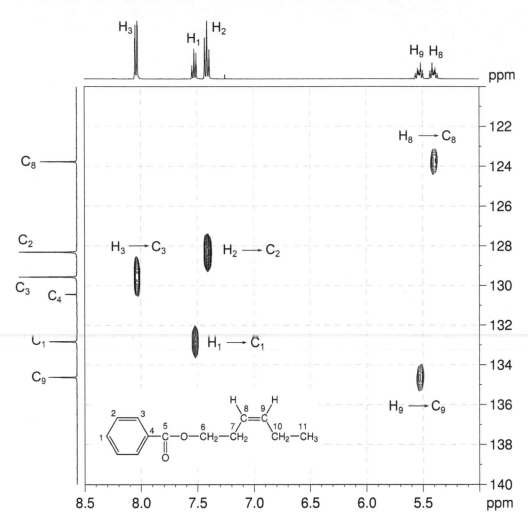

9. In the HMBC spectrum, the aromatic proton to carbonyl correlations (H_3–C_5 and H_2–C_5) and the –OCH_2– proton to carbonyl correlation (H_6–C_5) place the carbonyl group between the phenyl ring and the –OCH_2– group to afford the compound as $PhC(=O)OCH_2CH_2CH=CHCH_2CH_3$.

10. Note that the weak correlation between H_2 and C_5 is a long-range four-bond coupling.

^1H–^{13}C HMBC spectrum of *cis*-3-hexenyl benzoate (CDCl₃, 400 MHz)

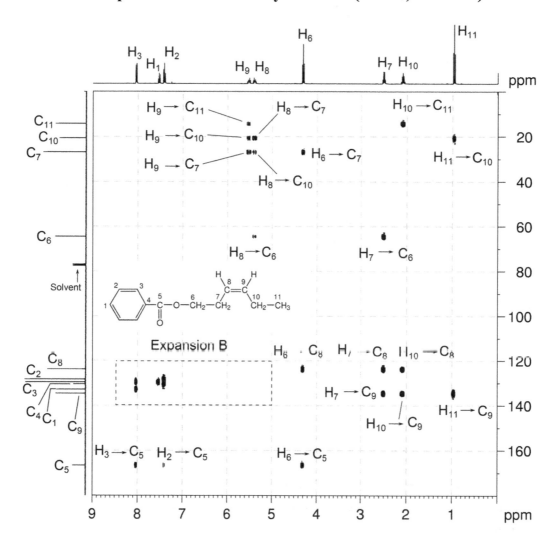

11. In the HMBC spectrum, the correlation from the aromatic proton signal at 7.41 ppm (H_2) to the quaternary aromatic carbon signal at 130.5 ppm confirms the assignment of the signal at 130.5 ppm as due to the *ipso* carbon C_4.

12. All other correlations in the HMBC spectrum are consistent with the structure.

^1H–^{13}C HMBC spectrum of *cis*-3-hexenyl benzoate – expansion B

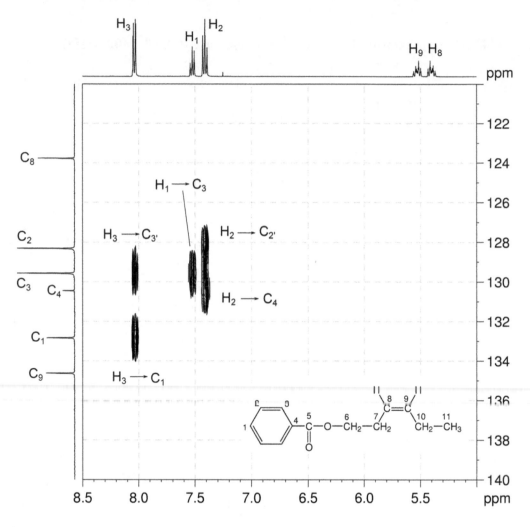

13. In the HMBC spectrum, the H_3–$C_{3'}$ and H_2–$C_{2'}$ correlations, while appearing to be one-bond correlations, actually arise from the $^3J_{C-H}$ interaction of a proton with the carbon *meta* to it.

14. The geometry about the double bond can be deduced from the NOESY spectrum. In particular the H_7–H_{10} correlation places the two –CH_2– groups on the same side of the double bond and thus a *cis* geometry can be deduced.

^1H–^1H NOESY spectrum of *cis*-3-hexenyl benzoate (CDCl$_3$, 400 MHz)

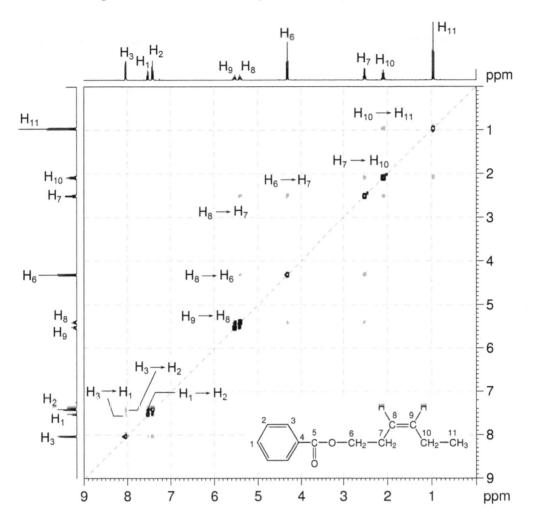

Problem 31

Question:

Identify the following compound.

Molecular Formula: $C_9H_{14}O$

IR: 1693 cm^{-1}

Solution:

trans-2,cis-6-Nonadienal

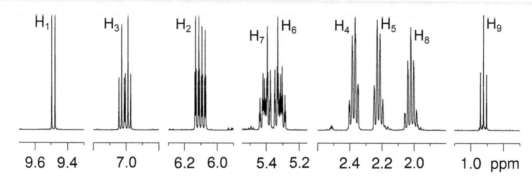

1. The molecular formula is $C_9H_{14}O$. Calculate the degree of unsaturation from the molecular formula: ignore the O atom to give an effective molecular formula of C_9H_{14} (C_nH_m) which gives the degree of unsaturation as $(n - m/2 + 1) = 9 - 7 + 1 = 3$. The compound contains a combined total of three rings and/or π bonds.

^{1}H NMR spectrum of *trans*-2,*cis*-6-nonadienal (DMSO-d_6, 400 MHz)

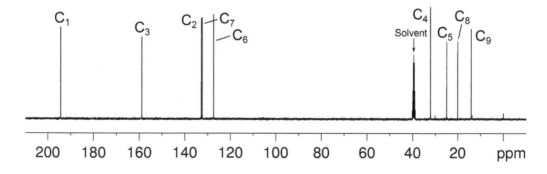

^{13}C{^{1}H} NMR spectrum of *trans*-2,*cis*-6-nonadienal (DMSO-d_6, 100 MHz)

2. In the HSQC spectrum, the ^1H 9.49 ppm / ^{13}C 194.3 ppm correlation (H$_1$–C$_1$) confirms the presence of an aldehyde group.

3. In the HSQC spectrum, the ^1H 7.01 ppm / ^{13}C 158.8 ppm (H$_3$–C$_3$), ^1H 6.10 ppm / ^{13}C 132.8 ppm (H$_2$–C$_2$), ^1H 5.40 ppm / ^{13}C 132.4 ppm (H$_7$–C$_7$) and ^1H 5.32 ppm / ^{13}C 127.4 ppm (H$_6$–C$_6$) correlations indicate the presence of two alkene groups.

4. In the HSQC spectrum, the ^1H 2.38 ppm / ^{13}C 32.2 ppm (H$_4$–C$_4$), ^1H 2.22 ppm / ^{13}C 25.0 ppm (H$_5$–C$_5$) and ^1H 2.02 ppm / ^{13}C 20.1 ppm (H$_8$–C$_8$) correlations indicate the presence of three –CH$_2$– groups whilst the ^1H 0.92 ppm / ^{13}C 14.1 ppm (H$_9$–C$_9$) correlation indicates the presence of a methyl group.

5. The aldehyde and two alkene functional groups account for all of the degrees of unsaturation, so the compound contains no additional rings or multiple bonds.

^1H–^{13}C me-HSQC spectrum of *trans*-2,*cis*-6-nonadienal (DMSO-d_6, 400 MHz)

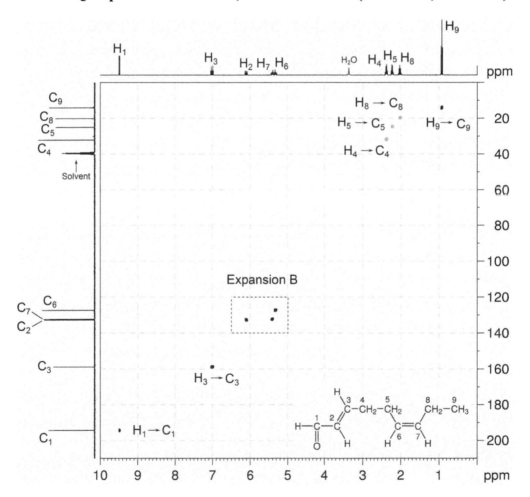

$^1H-^{13}C$ me-HSQC spectrum of *trans*-2,*cis*-6-nonadienal – expansion B

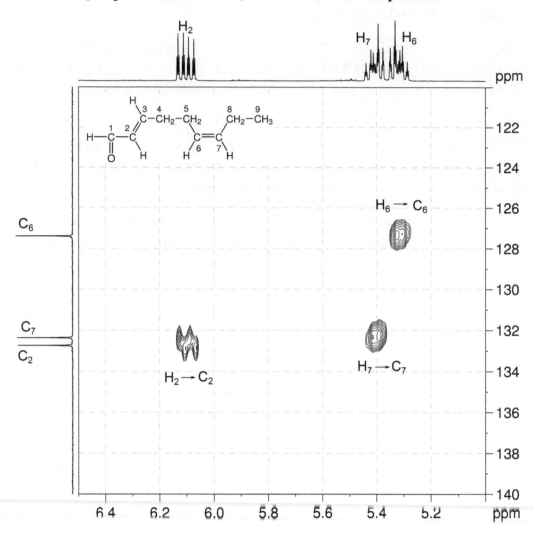

6. In the COSY spectrum, the H_1–H_2, H_2–H_3, H_3–H_4 and H_5–H_6 correlations afford the HC(O)–CH=CH–CH$_2$–CH$_2$–CH= fragment. The H_8–H_9 and H_7–H_8 correlations afford the CH$_3$–CH$_2$–CH= fragment.

¹H–¹H COSY spectrum of *trans*-2,*cis*-6-nonadienal (DMSO-*d*₆, 400 MHz)

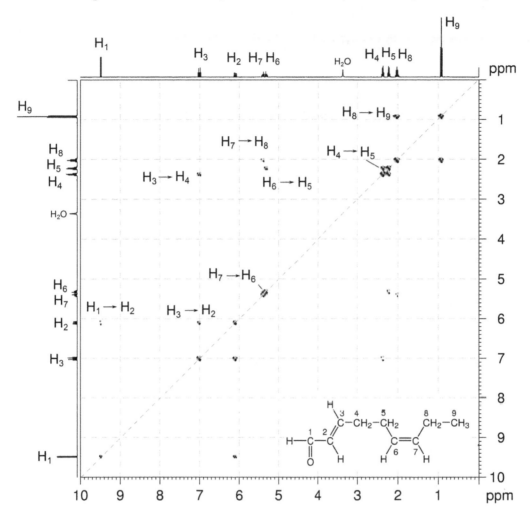

7. All of the nuclei in the molecular formula have been accounted for. The two fragments can be linked together in only one way to afford HC(O)CH=CHCH$_2$CH$_2$CH=CHCH$_2$CH$_3$.

8. The HMBC spectrum can be used to confirm the linkage of the two fragments. This is shown by the $H_6 \rightarrow C_8$ and $H_7 \rightarrow C_5$ correlations in expansion C and the $H_5 \rightarrow C_7$ and $H_8 \rightarrow C_6$ correlations in expansion E.

9. Note that the one-bond coupling between H_1 and C_1 is visible in the HMBC spectrum as a large doublet.

10. There is also a long-range four-bond correlation between H_4 and C_1.

11. All other correlations in the HMBC spectrum are consistent with the structure.

^1H–^{13}C HMBC spectrum of *trans*-2,*cis*-6-nonadienal (DMSO-*d_6*, 400 MHz)

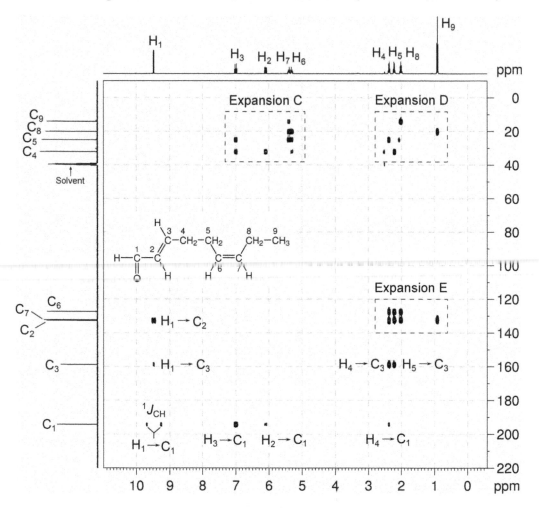

^{1}H–^{13}C HMBC spectrum of *trans*-2,*cis*-6-nonadienal – expansion C

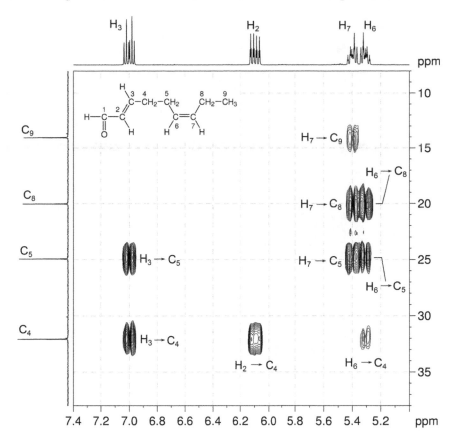

^{1}H–^{13}C HMBC spectrum of *trans*-2,*cis*-6-nonadienal – expansion D

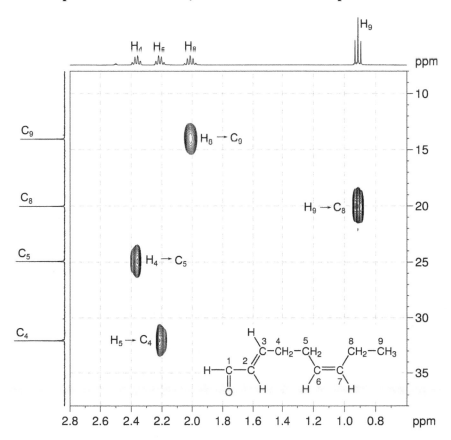

^{1}H–^{13}C HMBC spectrum of *trans*-2,*cis*-6-nonadienal – expansion E

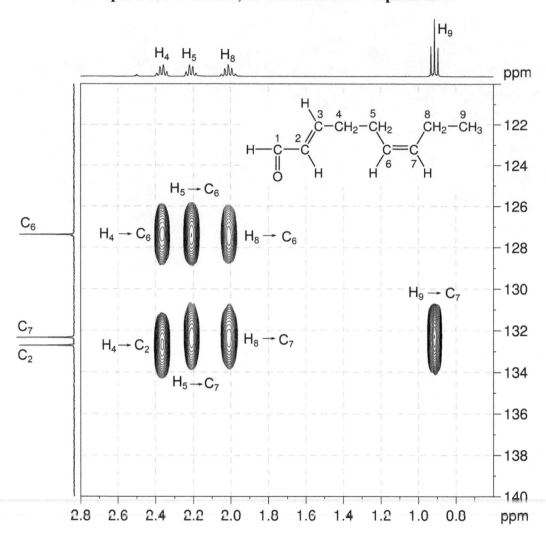

12. The stereochemistry about the double bonds can be determined from the NOESY spectrum.

13. In the NOESY spectrum, the H_1–H_3, H_2–H_4 and H_3–H_4 correlations indicate a *trans* geometry about the double bond adjacent to the aldehyde group.

14. The geometry about the second alkene group is determined particularly by the H_5–H_8 correlation which places the two –CH_2– groups on the same side of the double bond, *i.e.* a *cis* geometry.

^{1}H–^{1}H NOESY spectrum of *trans*-2,*cis*-6-nonadienal (CDCl$_3$, 400 MHz)

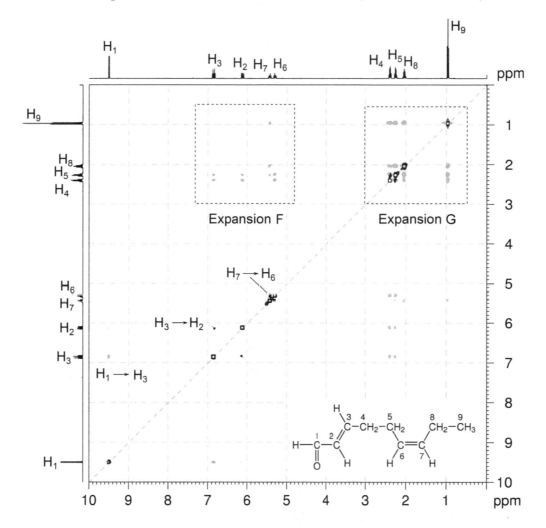

^{1}H–^{1}H NOESY spectrum of *trans*-2,*cis*-6-nonadienal – expansion F

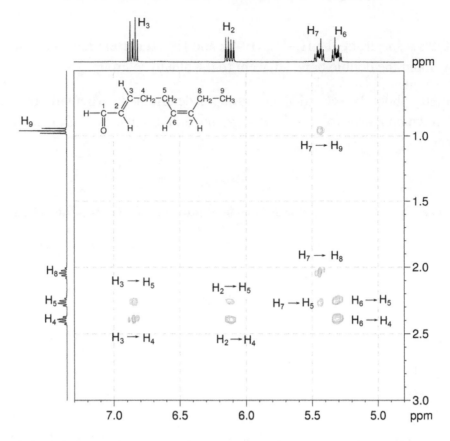

^{1}H–^{1}H NOESY spectrum of *trans*-2,*cis*-6-nonadienal – expansion G

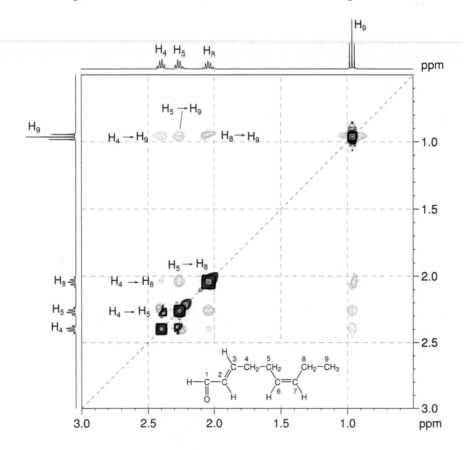

Question:

Identify the following compound.

Molecular Formula: $C_6H_{10}O_2$

IR Spectrum: 1649 (w), 1097 cm^{-1}

Solution:

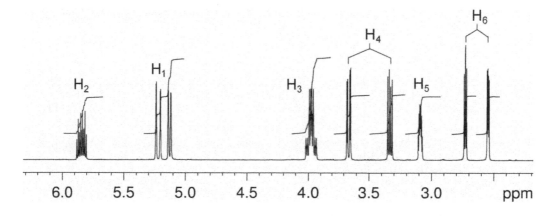

Allyl glycidyl ether

1. The molecular formula is $C_6H_{10}O_2$. Calculate the degree of unsaturation from the molecular formula: ignore the O atoms to give an effective molecular formula of C_6H_{10} (C_nH_m) which gives the degree of unsaturation as $(n - m/2 + 1) = 6 - 5 + 1 = 2$. The compound contains a combined total of two rings and/or π bonds.

2. The IR and $^{13}C\{^1H\}$ NMR spectra eliminate the possibility of a carbonyl-containing functional group.

3. There are nine resonances in the ^1H NMR spectrum, and six resonances in the $^{13}C\{^1H\}$ NMR spectrum. Some groups of protons must therefore be diastereotopic, so it is likely that either a ring system or a chiral centre is present in the molecule.

^1H NMR spectrum of allyl glycidyl ether (CDCl$_3$, 500 MHz)

$^{13}C\{^1H\}$ NMR spectrum of allyl glycidyl ether (CDCl$_3$, 125 MHz)

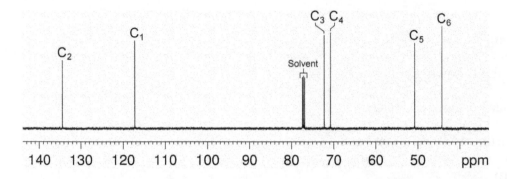

4. The me-HSQC spectrum easily identifies the protons (including diastereotopic pairs) and their associated carbon resonances: 5.84 ppm with 134.4 ppm (CH), 5.22 and 5.12 ppm with 117.2 ppm (CH$_2$), 3.98 ppm with 72.3 ppm (CH$_2$), 3.66 and 3.33 ppm with 70.8 ppm (CH$_2$), 3.09 ppm with 50.7 ppm (CH), and 2.73 and 2.55 ppm with 44.3 ppm (CH$_2$).

5. The chemical shifts of the carbon resonances at 134.4 and 117.2 ppm indicate the presence of a double bond, while the three protons associated with these signals show that the double bond is mono-substituted (CH$_2$=CHX).

^1H–^{13}C me-HSQC spectrum of allyl glycidyl ether (CDCl$_3$, 500 MHz)

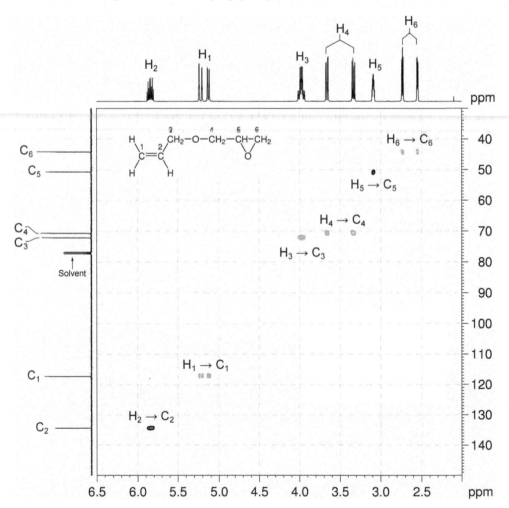

6. The $^1H–^1H$ COSY spectrum identifies two isolated spin systems in the molecule. The first involves the alkene signals (at 5.84 ppm, and 5.22/5.12 ppm) and the methylene signal at 3.98 ppm. The alkene-based spin system may be expanded to $CH_2=CHCH_2X$.

7. The second spin system identified by the COSY spectrum involves the remaining resonances in the ^1H NMR spectrum. The methylene resonances at 3.66 and 3.33 ppm correlate to the CH resonance at 3.09 ppm, which further correlates to the methylene signals at 2.73 and 2.55 ppm. This spin system therefore consists of a $–CH_2CHCH_2–$ fragment.

$^1H–^1H$ COSY spectrum of allyl glycidyl ether (CDCl$_3$, 500 MHz)

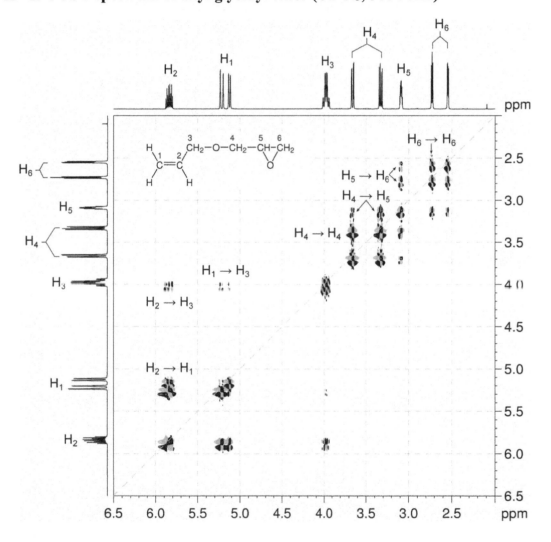

8. We have accounted for all the ^1H and ^{13}C nuclei in the formula; however two oxygen atoms remain unaccounted for. The two unique spin systems must be linked by one oxygen atom, so we have an ether: $CH_2=CHCH_2OCH_2CHCH_2$.

9. The second oxygen atom must "cap" the terminal CH and CH₂ groups, so each carbon atom forms the expected four bonds. The compound is therefore:

$$H_2C=CH-CH_2-O-CH_2-CH-CH_2$$

10. The HMBC spectrum shows expected two- and three-bond CH correlations (with the exception of $H_5 \rightarrow C_6$), including those that occur across the bridging oxygen (*i.e.* $H_3 \rightarrow C_4$ and $H_4 \rightarrow C_3$).

^1H–^{13}C HMBC spectrum of allyl glycidyl ether (CDCl₃, 500 MHz)

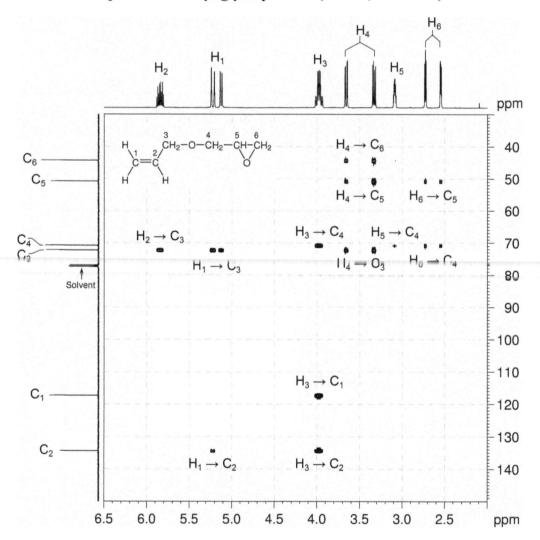

Question:

Identify the following compound.

Molecular Formula: $C_6H_{10}O_2$

IR: 1712 cm^{-1}

Solution:

3,4-Epoxy-4-methyl-2-pentanone

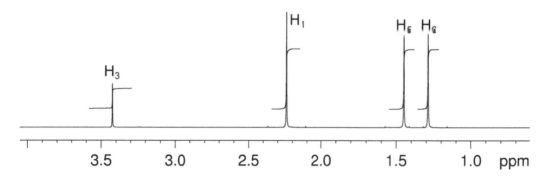

1. The molecular formula is $C_6H_{10}O_2$. Calculate the degree of unsaturation from the molecular formula: ignore the O atoms to give an effective molecular formula of C_6H_{10} (C_nH_m) which gives the degree of unsaturation as $(n - m/2 + 1) = 6 - 5 + 1 = 2$. The compound contains a combined total of two rings and/or π bonds.

2. The ^{1}H NMR spectrum shows three isolated methyl groups, and an isolated –CH group. The chemical shift of the latter suggests it is bound to an oxygen atom.

^{1}H NMR spectrum of 3,4-epoxy-4-methyl-2-pentanone (CDCl$_3$, 500 MHz)

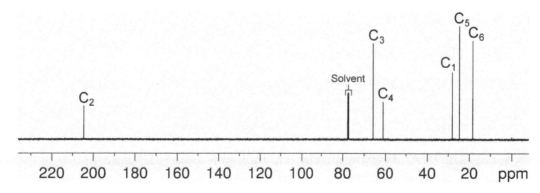

^{13}C{^{1}H} NMR spectrum of 3,4-epoxy-4-methyl-2-pentanone (CDCl$_3$, 125 MHz)

3. The ^{13}C{^{1}H} NMR spectrum shows a peak to low-field (204.6 ppm), consistent with the

presence of a ketone functional group.

4. The me-HSQC spectrum identifies the protonated carbon atoms at 65.7 for the OCH carbon, and 28.1, 24.7 and 18.4 ppm for the three methyl carbons. The chemical shift of the quaternary carbon signal at 61.0 ppm suggests it is bound directly to an oxygen atom.

^{1}H–^{13}C me-HSQC spectrum of 3,4-epoxy-4-methyl-2-pentanone (CDCl$_3$, 500 MHz)

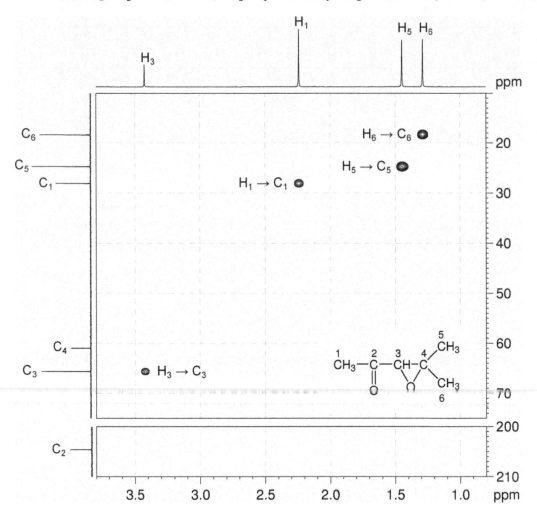

5. The structural elements identified so far are:

    a. A ketone (–C(=O)–).

    b. Three isolated methyl groups (3 × –CH$_3$).

    c. An isolated –CH– group bound to oxygen.

    d. A quaternary carbon atom bound to oxygen.

    e. The latter oxygen must be common to (c) and (d), suggesting either an ether or an epoxide.

6. Using the ^1H–^{13}C HMBC spectrum, the elements can be pieced together.

7. Starting with the ketone resonance at 204 ppm (C_2, which is arguably the most easily identified carbon resonance), there are correlations to the CH group (H_3) and one methyl group (H_1), giving us CH_3–C(=O)–CH–. H_1 also correlates to C_3. In this case, H_3 does not show a strong correlation to C_1 due to a Karplus-type bond-angle effect.

8. In addition to C_2, the CH proton H_3 also correlates to the quaternary signal at 60 ppm (C_4) and a second methyl group (C_5). The methyl group *cannot* be directly bound to the CH fragment (otherwise the protons would couple), so we now have CH_3–C(=O)–CH–C–CH_3. The possibility of an ether linkage between the CH (C_3) and CH_3 (C_5) groups can be ruled out based on the chemical shift of C_5.

9. The correlation from the protons of the remaining methyl group H_6 to quaternary carbon C_4 as well as the correlations of the two methyl groups to each other ($H_6 \rightarrow C_5$ and $H_5 \rightarrow C_6$) affords CH_3–C(=O)–CH–C–$(CH_3)_2$. The downfield chemical shifts of C_4 and C_3 indicate that they are bound to oxygen and there is only one oxygen atom left in the molecular formula so the compound must be an epoxide.

^1H–^{13}C HMBC spectrum of 3,4-epoxy-4-methyl-2-pentanone (CDCl$_3$, 500 MHz)

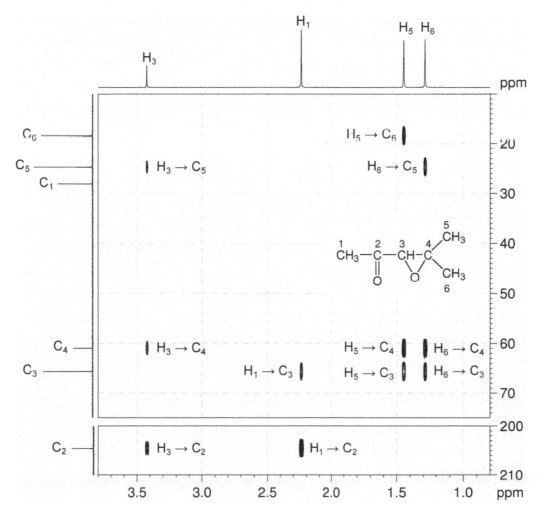

Problem 34

Question:

Given below are six compounds which are isomers of $C_5H_{11}NO_2S$. The ^1H and $^{13}C\{^1H\}$ NMR spectra of one of the isomers are given below. The ^1H–^{13}C me-HSQC and ^1H–^{13}C HMBC spectra are given on the following page. To which of these compounds do the spectra belong?

$$CH_3-S-CH_2-CH_2-\underset{\underset{H}{|}}{C}-\underset{\underset{O}{\|}}{C}-OH \qquad CH_3-S-CH_2-\underset{\underset{H}{|}}{C}-CH_2-\underset{\underset{O}{\|}}{C}-OH \qquad CH_3-S-\underset{\underset{H}{|}}{C}-CH_2-CH_2-\underset{\underset{O}{\|}}{C}-OH$$

$$\qquad\qquad NH_2 \qquad\qquad\qquad\qquad NH_2 \qquad\qquad\qquad\qquad NH_2$$

A   **B**   **C**

$$HS-CH_2-CH_2-\underset{\underset{H}{|}}{C}-\underset{\underset{O}{\|}}{C}-OCH_3 \qquad HS-CH_2-\underset{\underset{H}{|}}{C}-CH_2-\underset{\underset{O}{\|}}{C}-OCH_3 \qquad HS-\underset{\underset{H}{|}}{C}-CH_2-CH_2-\underset{\underset{O}{\|}}{C}-OCH_3$$

D   **E**   **F**

Solution:

dl-Methionine

(Isomer A)

$$\underset{CH_3}{\overset{1}{}}-S-\underset{CH_2}{\overset{2}{}}-\underset{CH_2}{\overset{3}{}}-\underset{4}{C}-\underset{5}{C}-OH$$
(with NH₂ on C4, H on C4, O on C5)

^{1}H NMR spectrum of *dl*-methionine (D₂O, 600 MHz)

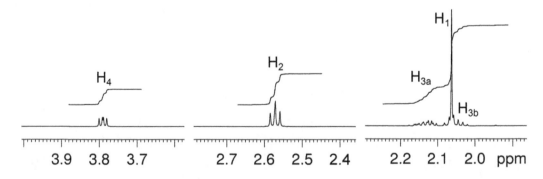

$^{13}C\{^1H\}$ NMR spectrum of *dl*-methionine (D_2O, 150 MHz)

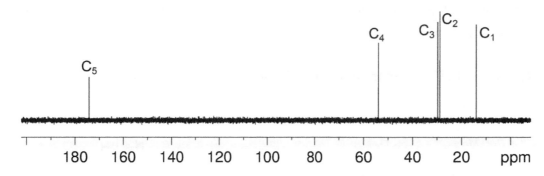

1. From the me-HSQC spectrum, the ^1H and ^{13}C chemical shifts of the methyl group H_1 and C_1 show that it is not bound to an oxygen atom. This excludes isomers D, E and F which are methyl esters.

2. From the coupling patterns of the two $-CH_2-$ and $-CH$ groups, both $-CH_2-$ groups are adjacent to each other *i.e.* $-CH_2CH_2CH-$ which excludes isomer B.

3. Note that the H_3 protons are diastereotopic as they are parts of a $-CH_2-$ group adjacent to a chiral carbon C_4.

^1H–^{13}C me-HSQC spectrum of *dl*-methionine (D_2O, 600 MHz)

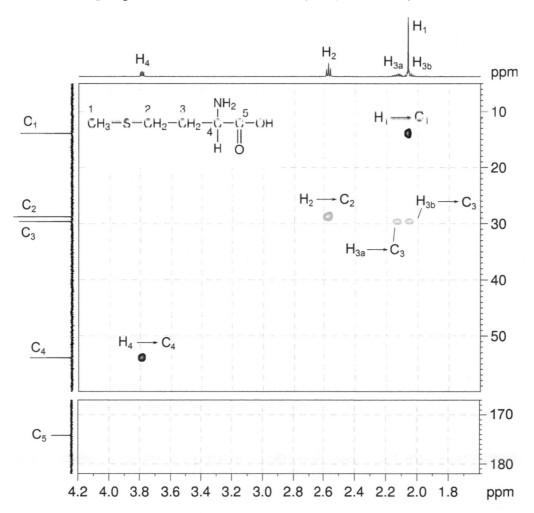

4. In the HMBC spectrum, the CH to carbonyl correlation (H$_4$–C$_5$) identifies the correct isomer as isomer A. In isomer C, this would be an unlikely four-bond correlation.

^{1}H–^{13}C HMBC spectrum of *dl*-methionine (D$_2$O, 600 MHz)

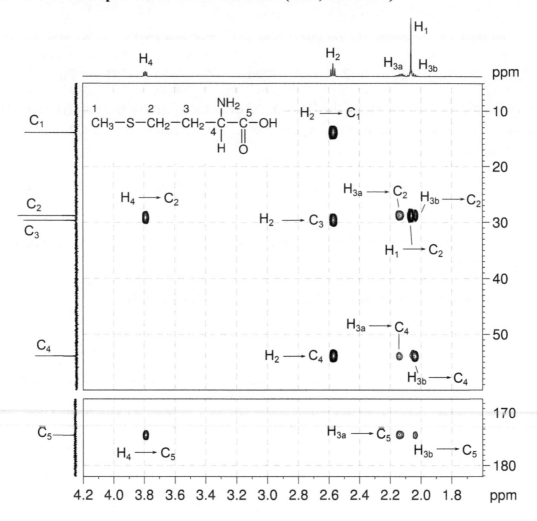

5. While this compound does contain a chiral centre, NMR cannot establish the absolute stereochemistry of the chiral centre.

Question:

Identify the following compound.

Molecular Formula: $C_8H_{15}NO_3$

IR: 3333, 1702, 1626 cm^{-1}

Solution:

N-Acetyl-*l*-leucine

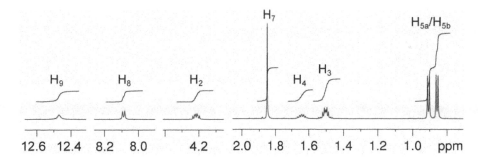

1. The molecular formula is $C_8H_{15}NO_3$. Calculate the degree of unsaturation from the molecular formula: ignore the O atoms, ignore the N atom and remove one H to give an effective molecular formula of C_8H_{14} (C_nH_m) which gives the degree of unsaturation as $(n - m/2 + 1) = 8 - 7 + 1 = 2$. The compound contains a combined total of two rings and/or π bonds.

2. IR and 1D spectra establish that there are two carbonyl groups (^{13}C resonances at 174.7 and 169.7 ppm). This accounts for all of the degrees of unsaturation, so the compound contains no additional rings or multiple bonds.

3. The ^{1}H resonance at 8.09 ppm (H_8) exchanges with D_2O on warming and indicates the presence of an amide group (O=C–NH). The ^{1}H resonance at 12.47 ppm (H_9) exchanges with D_2O and must be due to a carboxylic acid.

4. Further inspection of the ^{1}H NMR spectrum shows the presence of two –CH groups at 4.22 and 1.64 ppm (H_2 and H_4), one –CH$_2$– group at 1.50 ppm (H_3) and three –CH$_3$ groups at 1.85, 0.91 and 0.86 ppm (H_7, H_{5a} and H_{5b}). The multiplicities of these signals are confirmed by the correlations in the me-HSQC spectrum.

^{1}H NMR spectrum of *N*-acetyl-*l*-leucine (DMSO-d_6, 600 MHz)

$^{13}C\{^1H\}$ NMR spectrum of *N*-acetyl-*l*-leucine (DMSO-d_6, 150 MHz)

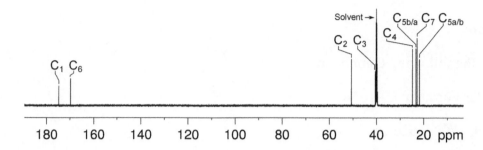

5. In the COSY spectrum, the amide resonance H_8 correlates with the CH resonance H_2. Further correlations between H_2–H_3, H_3–H_4 and H_4–H_{5a}/H_{5b} extend the skeleton to O=C–NH–CH–CH$_2$–CH(CH$_3$)$_2$.

6. Note that the H_3 protons are diastereotopic as they are part of a –CH$_2$– group adjacent to chiral centre C_2.

^1H–^1H COSY spectrum of *N*-acetyl leucine (DMSO-d_6, 600 MHz)

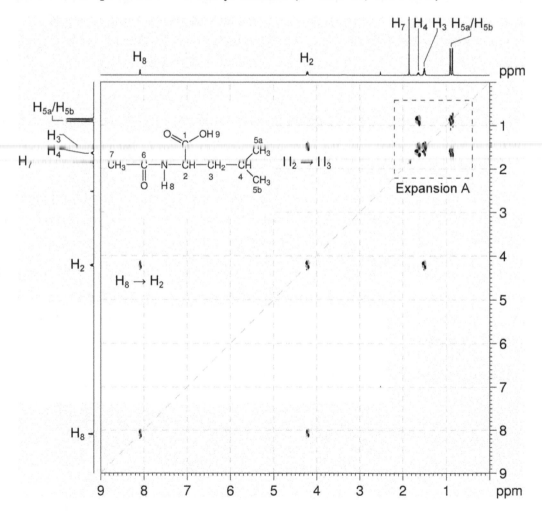

^{1}H–^{1}H COSY spectrum of *N*-acetyl leucine – expansion A

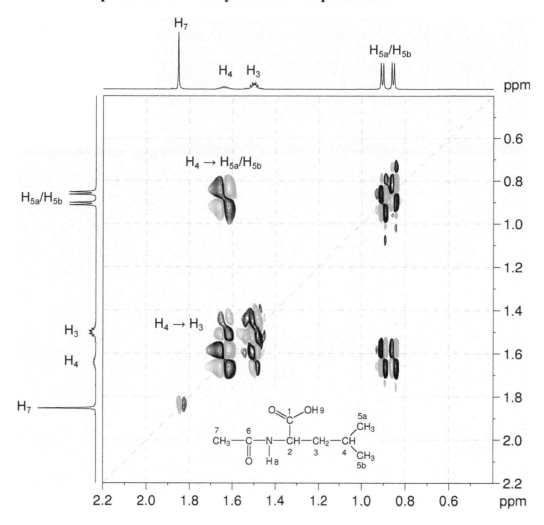

7. The me-HSQC spectrum easily identifies the ^{13}C resonances of the protonated carbons: the two CH carbons are at 50.7 and 24.8 ppm (C_2 and C_4), the CH_2 carbon is at 40.5 ppm (C_3), and the three methyl carbons are at 23.3, 21.8 and 22.8 ppm (C_{5a}, C_{5b} and C_7).

^1H–^{13}C me-HSQC spectrum of *N*-acetyl leucine (DMSO-d_6, 600 MHz)

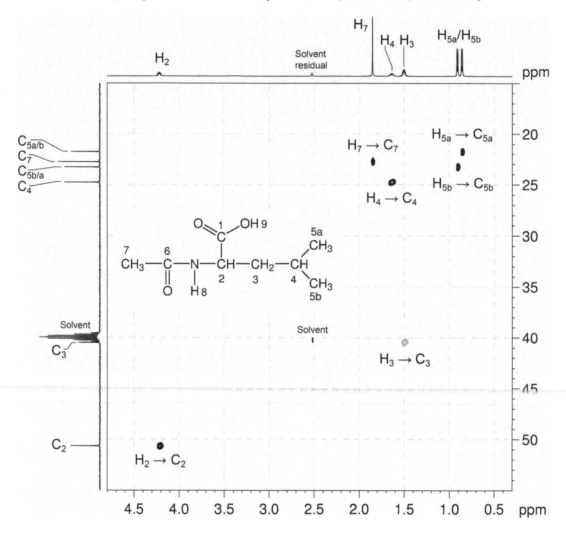

8. In the HMBC spectrum, the H_8–C_6 correlation identifies C_6 as the amide carbonyl.

9. The H_2–C_1 and H_3–C_1 correlations in the HMBC spectrum indicate that the carboxylic acid group C_1 is bound to C_2.

10. The remaining methyl group is located next to the amide by the H_7–C_6 correlation in the HMBC spectrum.

11. While this compound does contain a chiral centre, NMR cannot establish the absolute stereochemistry of the chiral centre.

¹H–¹³C HMBC spectrum of *N*-acetyl leucine (DMSO-*d₆*, 600 MHz)

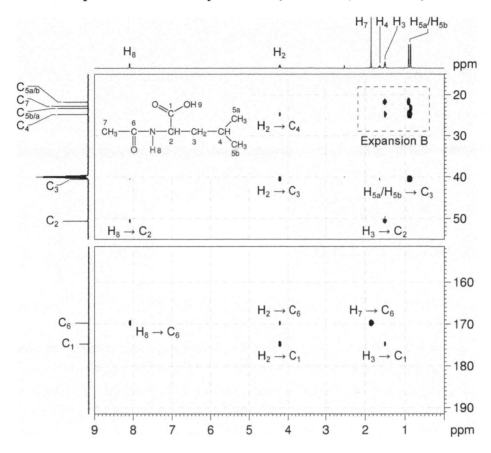

¹H–¹³C HMBC spectrum of *N*-acetyl leucine – expansion B

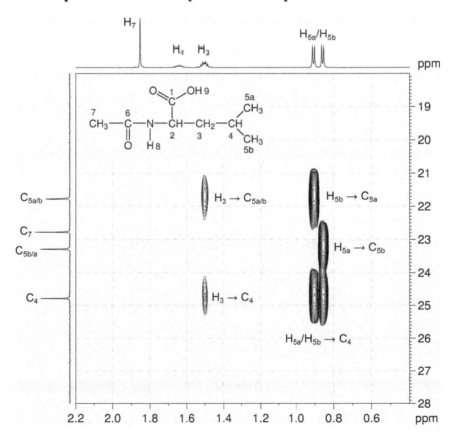

Problem 36

Question:

Identify the following compound.

Molecular Formula: $C_{10}H_{20}O_2$

IR: 1739 cm^{-1}

Solution:

Isoamyl valerate

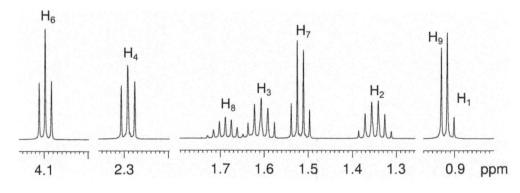

1. The molecular formula is $C_{10}H_{20}O_2$. Calculate the degree of unsaturation from the molecular formula: ignore the O atoms to give an effective molecular formula of $C_{10}H_{20}$ (C_nH_m) which gives the degree of unsaturation as $(n - m/2 + 1) = 10 - 10 + 1 = 1$. The compound contains one ring or one functional group containing a double bond.

2. IR and 1D spectra establish that the compound contains an ester functional group (^{13}C resonance at 174.0 ppm). This accounts for all of the degrees of unsaturation, so the compound contains no additional rings or multiple bonds.

3. Inspection of the ^1H NMR spectrum identifies five –CH$_2$– groups at 4.10, 2.29, 1.61, 1.52 and 1.35 ppm, one –CH group at 1.69 ppm and three overlapping –CH$_3$ groups at 0.92 ppm. The multiplicities of these signals are confirmed using the me-HSQC spectrum.

4. The ^1H signal at 4.10 ppm (H$_6$) must be bound to an oxygen atom based on chemical shift.

^1H NMR spectrum of isoamyl valerate (CDCl$_3$, 500 MHz)

$^{13}C\{^1H\}$ NMR spectrum of isoamyl valerate (CDCl$_3$, 125 MHz)

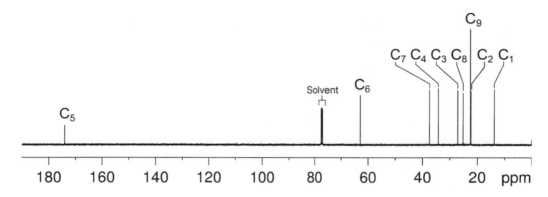

5. The TOCSY spectrum identifies all the correlated spins in a spin system. There are two spin systems – one containing three –CH$_2$– and one –CH$_3$ groups (H$_1$–H$_2$–H$_3$–H$_4$) and the other containing two –CH$_2$–, one –CH and two –CH$_3$ groups (H$_6$–H$_7$–H$_8$–H$_9$).

^{1}H–^{1}H TOCSY spectrum of isoamyl valerate (CDCl$_3$, 500 MHz)

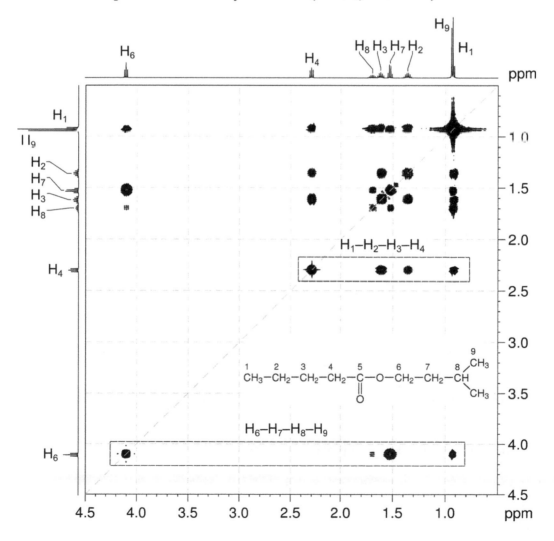

6. There is only one way to assemble the first spin system – $CH_2CH_2CH_2CH_3$.

7. There are two possible arrangements for the second spin system:

$$O-CH_2-CH_2-CH\begin{smallmatrix}CH_3\\[2pt]CH_3\end{smallmatrix} \qquad O-CH_2-CH\begin{smallmatrix}CH_3\\[2pt]CH_2-CH_3\end{smallmatrix}$$

8. Note that there are two different CH_3 environments at 22.5 ($2 \times C_9$) and 13.7 (C_1) ppm in the me-HSQC spectrum. One must belong to the first spin system (C_1) while the remaining two methyl groups must be equivalent (C_9) *i.e.* the $OCH_2CH_2CH(CH_3)_2$ fragment is correct.

^1H–^{13}C me-HSQC spectrum of isoamyl valerate (CDCl$_3$, 500 MHz)

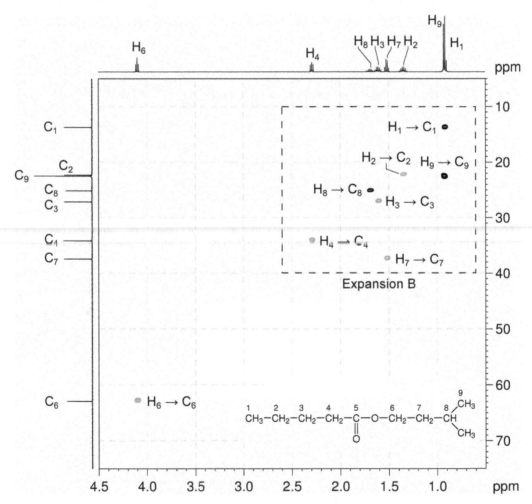

9. The molecule can be assembled in only one way.

Problem 37

Question:

The ^1H and $^{13}C\{^1H\}$ NMR spectra of (*E*)-4-methyl-4'-nitrostilbene ($C_{15}H_{13}NO_2$) recorded in acetone-d_6 solution at 298 K and 500 MHz are given below.

The ^1H NMR spectrum has signals at δ 2.34, 7.23, 7.33, 7.48, 7.56, 7.83 and 8.22 ppm.

The $^{13}C\{^1H\}$ NMR spectrum has signals at δ 21.3, 124.8, 126.2, 127.9, 128.0, 130.3, 134.1, 134.8, 139.5, 145.3 and 147.5 ppm.

The 2D ^1H–^1H COSY, multiplicity-edited ^1H–^{13}C HSQC and ^1H–^{13}C HMBC spectra are given on the following pages. Use these spectra to assign the ^1H and $^{13}C\{^1H\}$ resonances for this compound.

Solution:

(E)-4-Methyl-4'-nitrostilbene

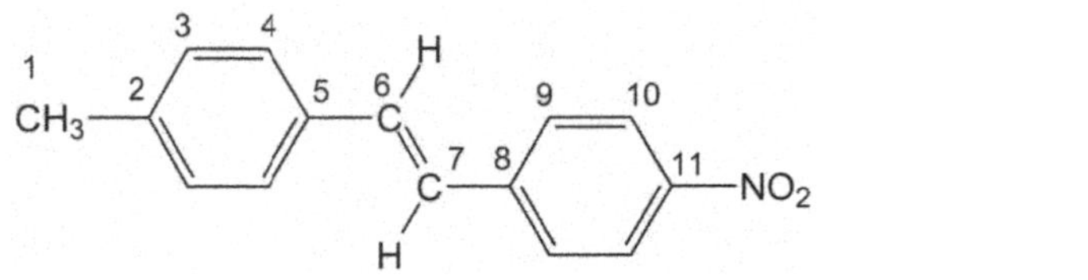

Proton	Chemical Shift (ppm)	Carbon	Chemical Shift (ppm)
H_1	2.34	C_1	21.3
		C_2	139.5
H_3	7.23	C_3	130.3
H_4	7.56	C_4	128.0
		C_5	134.8
H_6	7.48	C_6	134.1
H_7	7.33	C_7	126.2
		C_8	145.3
H_9	7.83	C_9	127.9
H_{10}	8.22	C_{10}	124.8
		C_{11}	147.5

1. The methyl group, H_1 and C_1, is easily identified in the ^1H and $^{13}C\{^1H\}$ NMR spectra, on the basis of its chemical shifts (2.34 and 21.3 ppm, respectively). The correlation between these two peaks in the me-HSQC confirms this assignment.

¹H NMR spectrum of (*E*)-4-methyl-4'-nitrostilbene (acetone-*d₆*, 500 MHz)

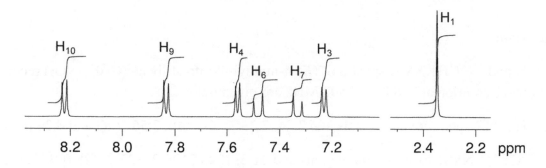

¹³C{¹H} NMR spectrum of (*E*)-4-methyl-4'-nitrostilbene (acetone-*d₆*, 125 MHz)

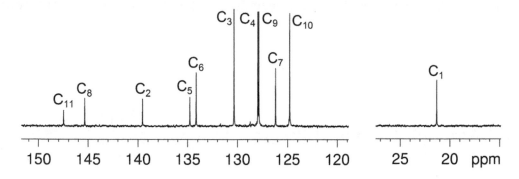

174

2. In expansion D of the HMBC spectrum, the methyl group (H_1) correlates to two signals at 130.3 and 139.5 ppm.

3. **Remember** that, in aromatic systems, the three-bond coupling $^3J_{C-H}$ is *typically* the larger long-range coupling and gives rise to the strongest cross-peaks.

^{1}H–^{13}C HMBC spectrum of (*E*)-4-methyl-4'-nitrostilbene – expansion D (acetone-*d*$_6$, 500 MHz)

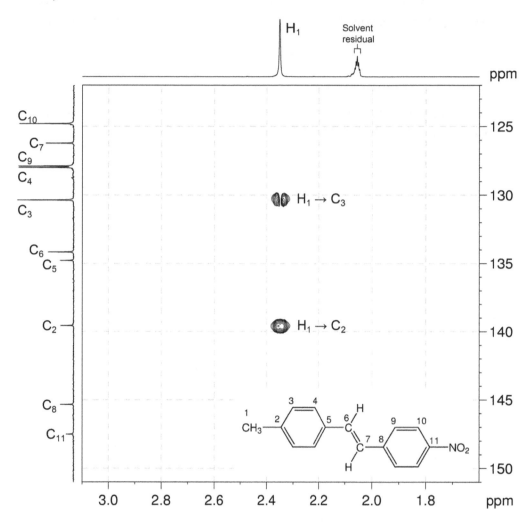

4. In the me-HSQC spectrum, the resonance at 130.3 ppm correlates to the ^{1}H signal at 7.23 ppm. This correlation identifies these signals as belonging to C_3 and H_3, respectively.

5. Further confirmation of the identity of H_3 can be seen in expansion B of the HMBC spectrum where there is a correlation from the ^{1}H signal at 7.23 ppm (H_3) to the methyl carbon (C_1).

6. In the me-HSQC spectrum, the resonance at 139.5 ppm does not correlate to any signals in the ^{1}H NMR spectrum. It is therefore a quaternary carbon (either C_2 or C_5). C_5 is too far removed from H_1 to experience coupling, so the signal at 139.5 ppm must arise from C_2.

^{1}H–^{13}C me-HSQC spectrum of (*E*)-4-methyl-4'-nitrostilbene – expansion A (acetone-d_6, 500 MHz)

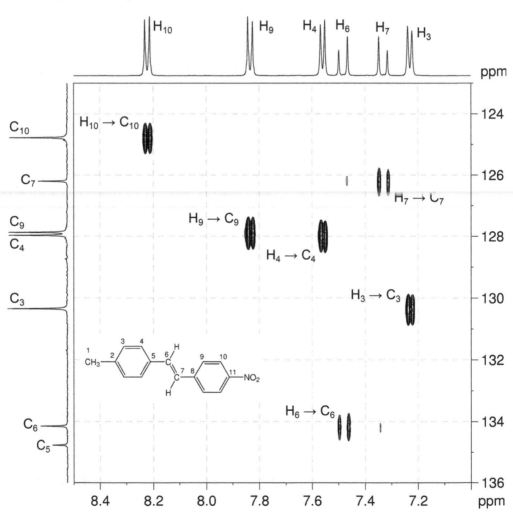

^{1}H–^{13}C HMBC spectrum of (*E*)-4-methyl-4'-nitrostilbene – expansion B

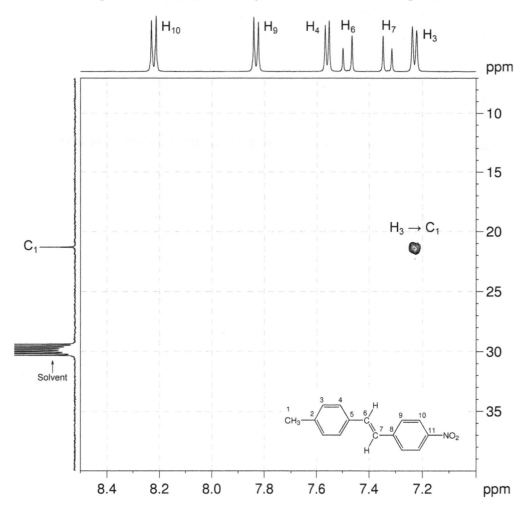

7. In expansion C of the HMBC spectrum, H_3 correlates to the ^{13}C signal at 134.8 ppm. Inspection of the me-HSQC shows that this signal arises from a quaternary carbon. H_3 is three bonds from C_5, which is quaternary. C_5 is therefore assigned.

$^1H–^{13}C$ HMBC spectrum of (*E*)-4-methyl-4'-nitrostilbene – expansion C

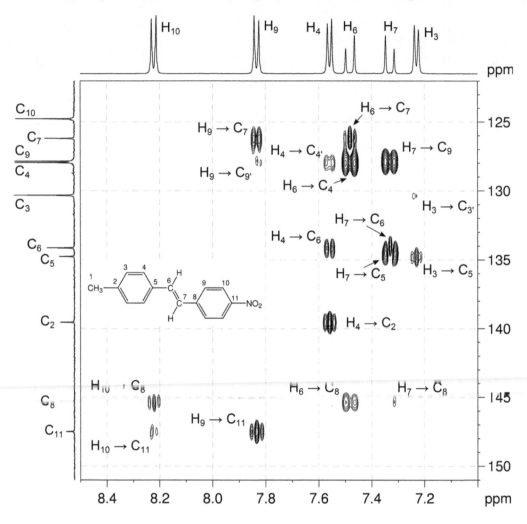

8. The aryl proton H_3 (at 7.23 ppm) correlates to the signal at 7.56 ppm in the COSY spectrum, identifying H_4.

^1H–^1H COSY spectrum of (*E*)-4-methyl-4'-nitrostilbene (acetone-d_6, 500 MHz)

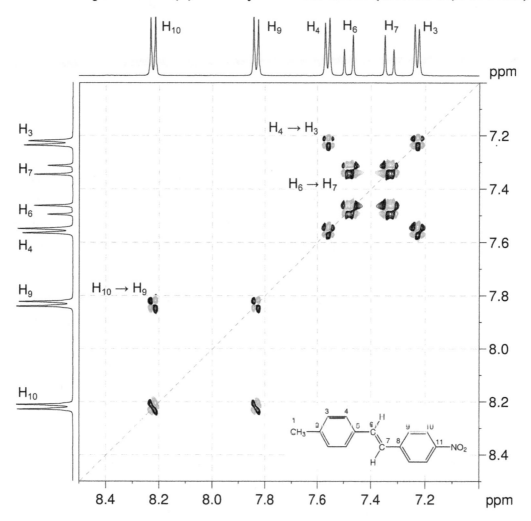

9. In the me-HSQC spectrum, H_4 (at 7.56 ppm) correlates to the ^{13}C signal at 128.0 ppm, identifying C_4.

10. In expansion C of the HMBC spectrum, H_4 correlates to three ^{13}C resonances. The first, to the signal at 128.0 ppm appears to be a one-bond correlation, however this signal arises from the $^3J_{C-H}$ interaction between the H_4 proton and the chemically equivalent $C_{4'}$ carbon which is three bonds away. The second correlation is to the signal at 134.1 ppm, while the third is the expected three-bond correlation to C_2 (139.5 ppm).

11. In the me-HSQC spectrum, the ^{13}C resonance at 134.1 ppm correlates to the ^1H signal at 7.48 ppm. These resonances are identified as belonging to C_6 and H_6 on the basis of the three bond separation between H_4 and C_6 (which gives rise to the correlation noted in the point above). In the ^1H NMR spectrum, the large splitting of this peak (16 Hz) confirms that the signal arises from one of the two vinyl protons.

12. In expansion C of the HMBC spectrum, H_6 correlates to three ^{13}C signals: the signal at 126.2 ppm, one of the two signals at ~128 ppm, and the signal at 145.3 ppm. H_6 is three bonds removed from C_4 (128.0 ppm) and C_8. The me-HSQC spectrum shows that the ^{13}C signal at 126.2 ppm correlates to the ^1H signal at 7.33, while the ^{13}C signal at 145.3 ppm is quaternary. C_8 (which is quaternary) is therefore assigned to the signal at 145.3 ppm. The signal at 126.2 ppm is not yet assigned.

13. In the COSY spectrum, H_6 correlates to the signal at 7.33 ppm, which identifies this signal as belonging to H_7. The me-HSQC spectrum then identifies C_7 (126.2 ppm). This identifies the unassigned correlation observed in the previous step.

14. In expansion C of the HMBC spectrum, H_7 correlates to three ^{13}C signals: one of the two signals at ~128 ppm, the signal at 134.1 ppm (C_6) and the signal at 134.8 (C_5). H_7 is three bonds removed from C_9, which is therefore identified as the signal at 127.9 ppm (as the signal at 128.0 ppm has been previously assigned to C_4). The me-HSQC spectrum identifies H_9 as the ^1H resonance at 7.83 ppm.

15. In expansion C of the HMBC spectrum, H_9 correlates strongly to two ^{13}C signals: at 126.2 (C_7) and 147.5 ppm. The me-HSQC spectrum shows that the resonance at 147.5 ppm belongs to a quaternary carbon. As H_9 is three bonds removed from C_{11}, this signal is assigned to C_{11}.

16. In the COSY spectrum, H_9 correlates to the signal at 8.22 ppm, which identifies this signal as belonging to H_{10}. The me-HSQC spectrum then identifies C_{10} (124.8 ppm).

Question:

Identify the following compound.

Molecular Formula: $C_{11}H_{16}O$

IR: 3450 (br) cm^{-1}

Solution:

2-*tert*-Butyl-6-methylphenol

1. The molecular formula is $C_{11}H_{16}O$. Calculate the degree of unsaturation from the molecular formula: ignore the O atom to give an effective molecular formula of $C_{11}H_{16}$ (C_nH_m) which gives the degree of unsaturation as $(n - m/2 + 1) = 11 - 8 + 1 = 4$. The compound contains a combined total of four rings and/or π bonds.

2. 1D spectra establish that the compound is a trisubstituted benzene. The aromatic ring accounts for all of the degrees of unsaturation, so the compound contains no additional rings or multiple bonds.

3. The coupling pattern in the expansion of the aromatic region of the ^{1}H NMR spectrum shows that the three protons on the aromatic ring must occupy adjacent positions. The proton at 6.78 ppm is the proton at the middle of the spin system since it has two large splittings (*i.e.* it has two *ortho*-protons) whereas the other two aromatic protons each have only one large splitting.

4. The three substituents on the aromatic ring must therefore be in adjacent positions *i.e.* this compound is a "1,2,3-trisubstituted benzene".

5. 1D spectra establish that the substituents are an –OH group (^{1}H resonance at 4.73 ppm, exchangeable), a –CH$_3$ group (three proton resonance at 2.22 ppm) and a *tert*-butyl group (nine proton resonance at 1.41 ppm) but, at this stage, it is not possible to establish which substituents are where.

^{1}H NMR spectrum of 2-*tert*-butyl-6-methylphenol (CDCl$_3$, 600 MHz)

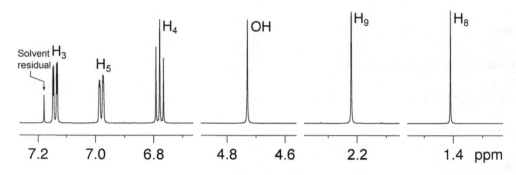

^{13}C{^{1}H} NMR spectrum of 2-*tert*-butyl-6-methylphenol (CDCl$_3$, 150 MHz)

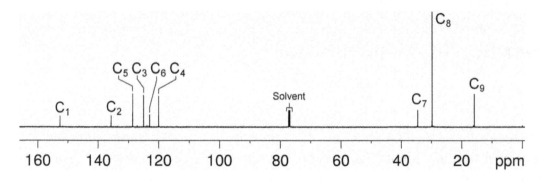

6. For a "1,2,3-trisubstituted benzene", there are three possible isomers:

A   B   C

7. The ¹H–¹³C me-HSQC spectrum easily identifies the protonated carbons. The three methyl groups of the *tert*-butyl group appear at 29.8 ppm and the aromatic methyl group is at 15.9 ppm.

8. The protonated aromatic carbons are at 120.0, 125.0 and 128.6 ppm and specifically the central proton at 6.78 ppm correlates to the carbon at 120.0 ppm. The remaining two aromatic protons at 7.14 and 6.98 ppm correlate to the carbons at 125.0 and 128.6 ppm, respectively.

9. The three aromatic carbons which bear the substituents appear at 152.7, 135.6 and 123.0 ppm. The low-field carbon resonance (152.7 ppm) must be the carbon bearing the –OH substituent based on its chemical shift.

¹H–¹³C me-HSQC spectrum of 2-*tert*-butyl-6-methylphenol (CDCl₃, 600 MHz)

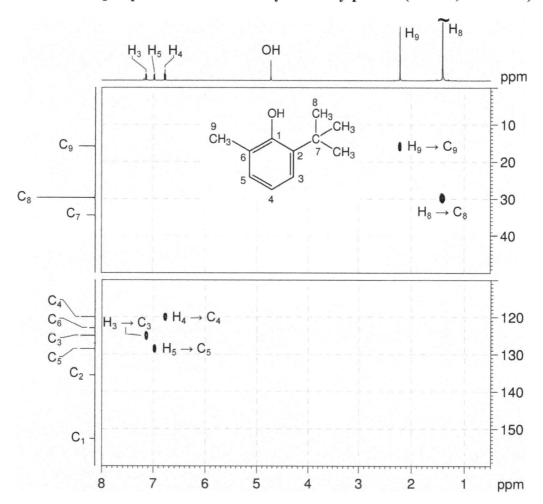

10. In the ^{1}H–^{13}C HMBC spectrum, the aromatic protons at the ends of the three-spin system (at 7.14 and 6.98 ppm) both correlate to the ^{13}C resonance at 152.7 ppm.

11. Remember that, in aromatic systems, the three-bond coupling $^3J_{\text{C–H}}$ is typically the larger long-range coupling and gives rise to the strongest cross-peaks. Therefore, the low-field carbon resonance (bearing the –OH substituent) must be *meta* to the protons at 7.14 and 6.98 ppm which identifies **Isomer C** as the correct answer.

^{1}H–^{13}C HMBC spectrum of 2-*tert*-butyl-6-methylphenol (CDCl$_3$, 600 MHz)

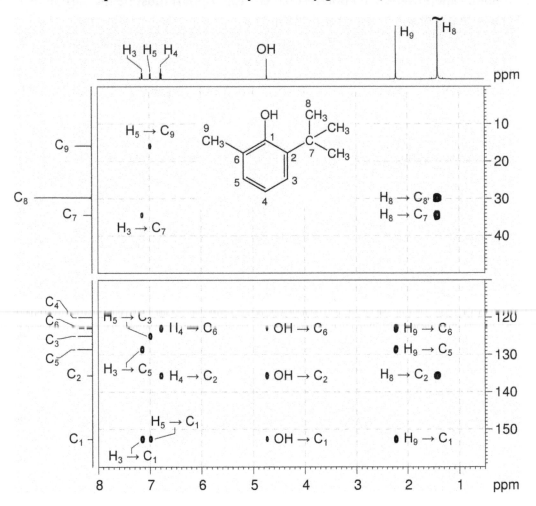

12. In the HMBC spectrum, the methyl protons (H$_9$) should correlate to C$_1$ and C$_5$ (three-bond correlations) as well as to the *ipso* carbon (C$_6$, a two-bond correlation). There is a correlation from H$_9$ to C$_1$ (at 153 ppm), as well as to the resonances at 128.6 (protonated) and 123.0 ppm (non-protonated). This identifies C$_5$ as the resonance at 128.6 ppm (and, from the HSQC spectrum, H$_5$ as the ^{1}H NMR resonance at 6.98 ppm), and C$_6$ as the resonance at 123.0 ppm. H$_3$ is therefore the resonance at 7.14 ppm, and C$_3$ the resonance at 125.0 ppm.

13. Similarly, the *tert*-butyl protons (H₈) should correlate to the *ipso* carbon (C₂, a three-bond correlation), and there is indeed a correlation between H₈ and the non-protonated carbon resonance at 135.6 ppm.

14. We would expect correlations to C₇ from H₃ ($^3J_{C-H}$) and H₈ ($^2J_{C-H}$), and these are present in the spectrum.

15. In the HMBC spectrum, the –OH proton also correlates to the *ipso* carbon C₁ (a two-bond correlation), and C₂ and C₆, which are three-bonds away. Note that correlations from exchangeable protons are not always observed.

16. Note also that in the HMBC spectrum for this compound, there is a strong correlation between H₈ and C₈′. While this appears to be a one-bond correlation, in *tert*-butyl groups, the apparent one-bond correlation arises from the $^3J_{C-H}$ interaction of the protons of one of the methyl groups with the chemically equivalent carbon which is three bonds away.

185

Question:

Identify the following compound.

Molecular Formula: $C_{10}H_{12}O$

IR: 3600 (br), 1638, 1594, 1469 (s) cm^{-1}

Solution:

2-Allyl-6-methylphenol

1. The molecular formula is $C_{10}H_{12}O$. Calculate the degree of unsaturation from the molecular formula: ignore the O atom to give an effective molecular formula of $C_{10}H_{12}$ (C_nH_m) which gives the degree of unsaturation as $(n - m/2 + 1) = 10 - 6 + 1 = 5$. The compound contains a combined total of five rings and/or π bonds.

2. 1D spectra establish that the compound is a trisubstituted benzene. The aromatic ring accounts for four degrees of unsaturation and the remaining degree of unsaturation is accounted for by a double bond (alkene ^1H resonances at 5.99 and 5.15 ppm).

3. The coupling pattern in the aromatic region of the ^1H NMR spectrum shows that the three protons on the aromatic ring must occupy adjacent positions. The proton at 6.77 ppm is the proton at the middle of the spin system since it has two large splittings (*i.e.* it has two *ortho*-protons) whereas the other two aromatic protons each have only one large splitting.

4. The three substituents on the aromatic ring must therefore be in adjacent positions *i.e.* this compound is a "1,2,3-trisubstituted benzene".

5. 1D spectra establish that two of the substituents are an –OH substituent (^1H resonance at 4.99 ppm, exchangeable) and a –CH_3 group (three proton resonance at 2.21 ppm).

^{1}H NMR spectrum of 2-allyl-6-methylphenol (CDCl$_3$, 500 MHz)

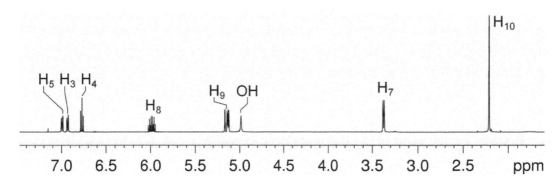

^{13}C{^{1}H} NMR spectrum of 2-allyl-6-methylphenol (CDCl$_3$, 125 MHz)

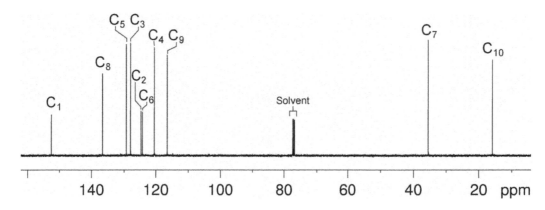

6. The ^{1}H–^{1}H COSY spectrum shows coupling from the resonance at 3.38 ppm (integration two protons) to the two resonances at 5.1 (CH$_2$) and 5.99 (CH) ppm. The chemical shift of the latter two resonances is consistent with alkene protons, so we can identify an allyl group (CH$_2$=CHCH$_2$–) as the third substituent on the aromatic ring. At this stage, it is not possible to establish which substituents are where.

^{1}H–^{1}H COSY spectrum of 2-allyl-6-methylphenol (CDCl$_3$, 500 MHz)

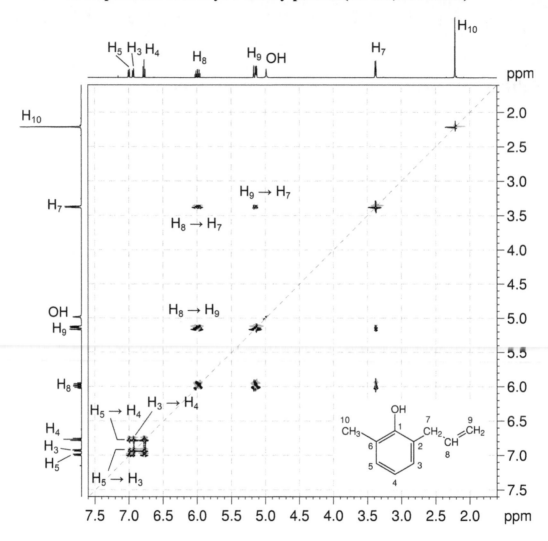

7. For a "1,2,3-trisubstituted benzene", there are three possible isomers:

A   B   C

8. The ^{1}H–^{13}C me-HSQC spectrum easily identifies the protonated carbons. The methyl group is at 15.8 ppm. The aliphatic CH_2 group is at 35.5 ppm while the alkene CH and CH_2 groups are at 136.7 and 116.6 ppm, respectively.

9. The protonated aromatic carbons are at 120.5, 128.1 and 129.4 ppm and specifically the central proton at 6.77 ppm correlates to the carbon at 120.5 ppm. The remaining two aromatic protons at 6.99 and 6.93 ppm correlate to the carbons at 129.4 and 128.0 ppm, respectively.

10. The three aromatic carbons which bear the substituents appear at 152.5, 124.7 and 124.2 ppm. The low-field carbon resonance (152.5 ppm) must be the carbon bearing the –OH substituent based on its chemical shift.

^{1}H–^{13}C me-HSQC spectrum of 2-allyl-6-methylphenol (CDCl$_3$, 500 MHz)

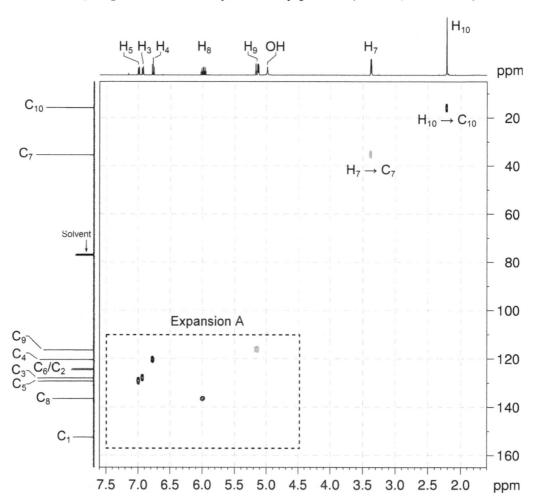

$^1H–^{13}C$ me-HSQC spectrum of 2-allyl-6-methylphenol – expansion A

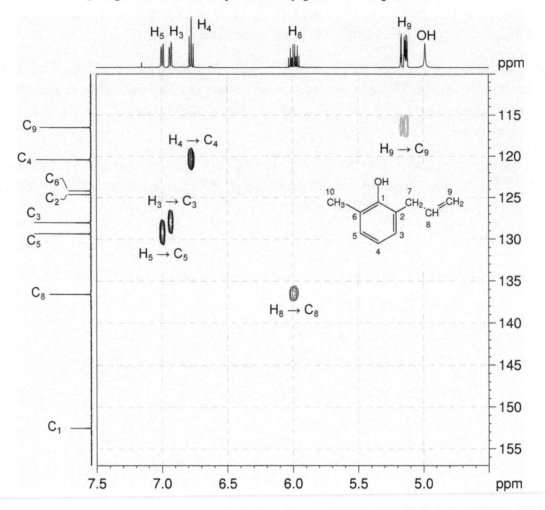

11. In the ^{1}H–^{13}C HMBC spectrum (expansion B), the aromatic protons at the ends of the three-spin system (at 6.99 and 6.93 ppm) both correlate to the ^{13}C resonance at 152.5 ppm.

12. Remember that, in aromatic systems, the three-bond coupling $^3J_{C-H}$ is typically the larger long-range coupling and gives rise to the strongest cross-peaks. Therefore, the low-field carbon resonance (bearing the –OH substituent) must be *meta* to the protons at 6.99 and 6.93 ppm which identifies **Isomer B** as the correct answer.

^{1}H–^{13}C HMBC spectrum of 2-allyl-6-methylphenol (CDCl$_3$, 500 MHz)

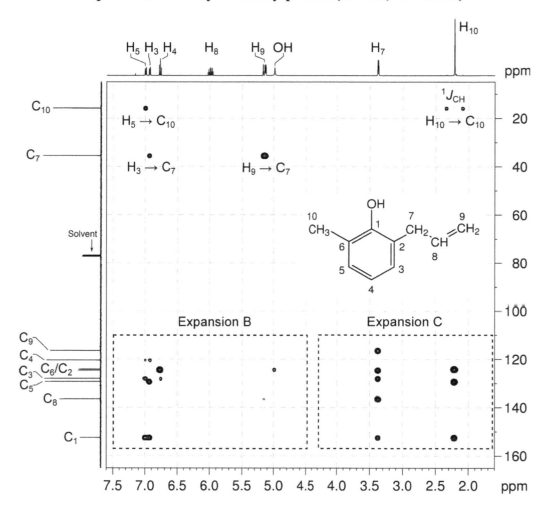

^{1}H–^{13}C HMBC spectrum of 2-allyl-6-methylphenol – expansion B

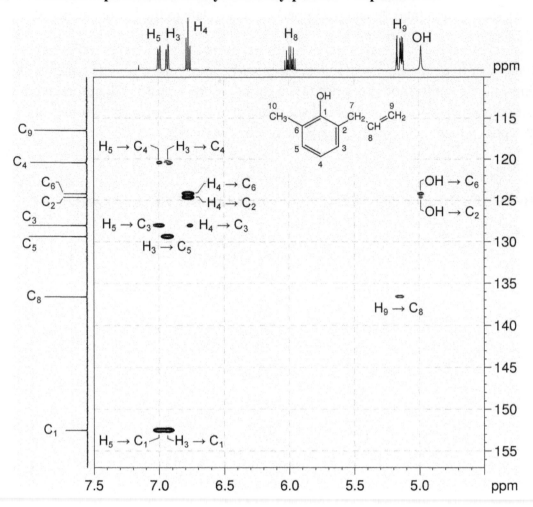

13. In the HMBC spectrum (expansion C), the methyl protons (H_{10}) should correlate to C_1 and C_5 (three-bond correlations) as well as to the *ipso* carbon (C_6, a two-bond correlation). There is a correlation from methyl proton H_{10} to C_1 (at 152.5 ppm), as well as to the resonances at 129.3 (protonated) and 124.2 ppm (non-protonated). This identifies C_5 as the resonance at 129.3 ppm (and, from the HSQC spectrum, H_5 as the ^1H NMR resonance at 6.99 ppm), and C_6 as the resonance at 124.2 ppm. H_3 is therefore the resonance at 6.93 ppm and C_3 the resonance at 128.0 ppm.

14. Similarly, the allyl protons alpha to the aromatic ring (H_7) should correlate to the *ipso* carbon (C_2, a two-bond correlation), and there is indeed a correlation between H_7 and the non-protonated carbon resonance at 124.7 ppm. We would also expect three-bond correlations to C_1 and C_3, as well as correlations to the rest of the allyl group, and these are all observed.

^1H–^{13}C HMBC spectrum of 2-allyl-6-methylphenol – expansion C

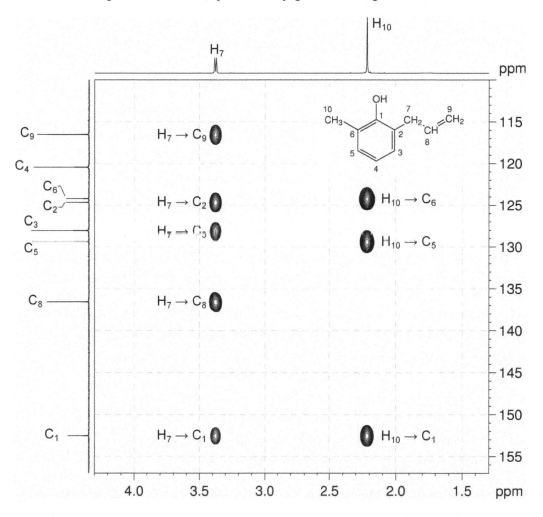

15. In the HMBC spectrum (expansion B), the –OH proton correlates to C_2 and C_6, which are three bonds away. The two-bond correlation to C_1 is not observed. Note that correlations from exchangeable protons are not always observed.

Question:

Identify the following compound.

Molecular Formula: $C_8H_8O_3$

IR: 3022 (br w), 1675, 1636 cm^{-1}

Solution:

2-Hydroxy-4-methoxybenzaldehyde

1. The molecular formula is $C_8H_8O_3$. Calculate the degree of unsaturation from the molecular formula: ignore the O atoms to give an effective molecular formula of C_8H_8 (C_nH_m) which gives the degree of unsaturation as $(n - m/2 + 1) = 8 - 4 + 1 = 5$. The compound contains a combined total of five rings and/or π bonds.

2. 1D NMR data establish that the compound is a trisubstituted benzene and contains an aldehyde functional group (δ_H 9.70 ppm, δ_C 194.4 ppm). This accounts for all of the degrees of unsaturation, so the compound contains no additional rings or multiple bonds.

3. The coupling pattern in the aromatic region of the ^{1}H NMR spectrum establishes that the substituents are in positions 1, 2 and 4 on the aromatic ring. The proton at 6.42 ppm clearly has no *ortho* couplings and must be the isolated proton in the spin system sandwiched between two substituents (H_3). The two protons at 6.53 and 7.42 ppm each have a large *ortho* coupling so they must be adjacent to each other. The proton at 6.53 ppm has an additional *meta* coupling so it must be H_5. The remaining proton at 7.42 ppm must be H_6.

4. From the 1D spectra, the substituents on the aromatic ring are an aldehyde, a methoxy group ($-OCH_3$, δ_H 3.85 ppm, δ_C 55.7 ppm) and a hydroxyl group ($-OH$, δ_H 11.5 ppm, exchangeable). It is not yet possible to establish which substituents are where.

^{1}H NMR spectrum of 2-hydroxy-4-methoxybenzaldehyde (CDCl$_3$, 600 MHz)

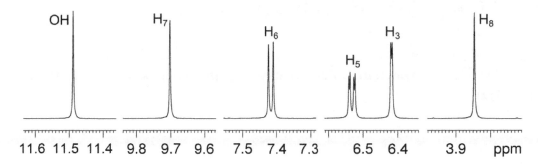

^{13}C{^{1}H} NMR spectrum of 2-hydroxy-4-methoxybenzaldehyde (CDCl$_3$, 150 MHz)

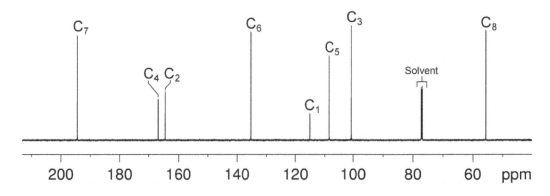

5. For a 1,2,4-trisubstituted aromatic ring, there are six possible isomers:

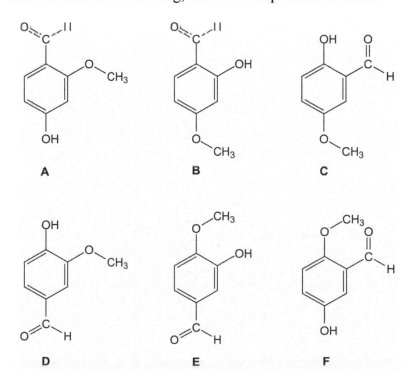

6. The ^1H–^{13}C me-HSQC spectrum easily identifies the protonated carbons: C_3 at 100.7, C_5 at 108.4, C_6 at 135.3, C_7 at 194.3 and C_8 at 55.7 ppm.

^1H–^{13}C me-HSQC spectrum of 2-hydroxy-4-methoxybenzaldehyde (CDCl$_3$, 600 MHz)

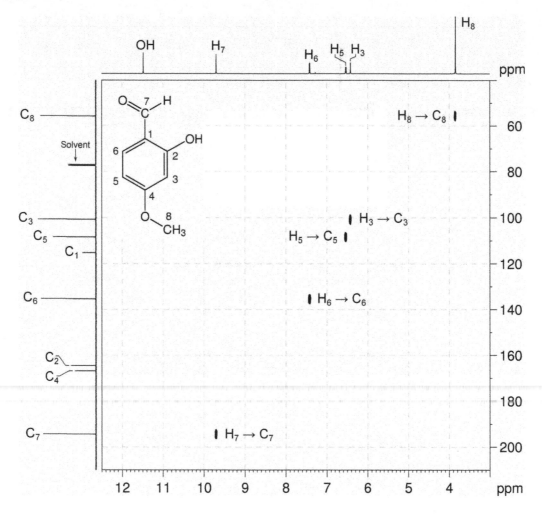

7. In the ^1H–^{13}C HMBC spectrum, the aromatic proton resonance H_6 correlates to the aldehyde carbon resonance C_7, thus placing the aldehyde group in position 1 of the ring.

8. The OH proton resonance correlates to aromatic carbon resonance C_3 in the HMBC spectrum thus the OH group is located adjacent to C_3 in position 2 of the benzene ring.

9. By elimination, the methoxy group is in position 4 of the benzene ring (**Isomer B**).

^1H–^{13}C HMBC spectrum of 2-hydroxy-4-methoxybenzaldehyde (CDCl₃, 600 MHz)

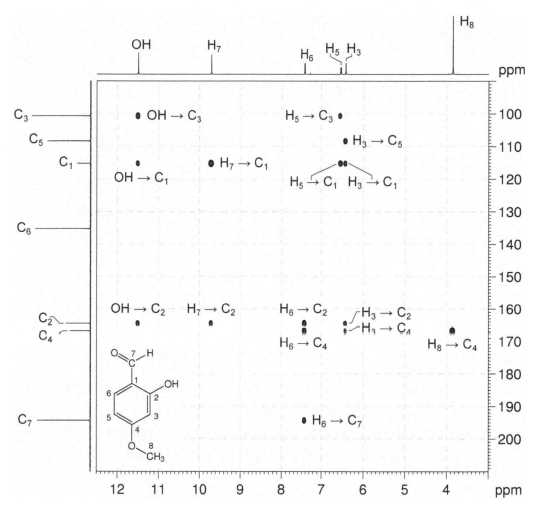

Question:

Identify the following compound.

Molecular Formula: $C_9H_{10}O_2$

IR: 1644, 1618 cm^{-1}

Solution:

2'-Hydroxy-5'-methylacetophenone

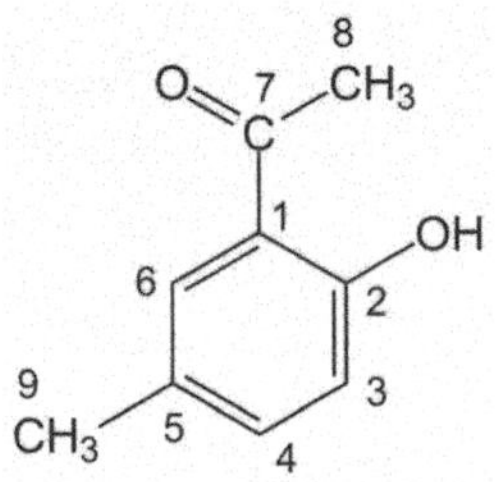

1. The molecular formula is $C_9H_{10}O_2$. Calculate the degree of unsaturation from the molecular formula: ignore the O atoms to give an effective molecular formula of C_9H_{10} (C_nH_m) which gives the degree of unsaturation as $(n - m/2 + 1) = 9 - 5 + 1 = 5$. The compound contains a combined total of five rings and/or π bonds.

2. IR and 1D NMR data establish the compound is a trisubstituted benzene and contains a ketone functional group (^{13}C resonance at 204.1 ppm). This accounts for all of the degrees of unsaturation, so the compound contains no additional rings or multiple bonds.

3. The coupling pattern in the expansion of the aromatic region of the ^1H NMR spectrum shows that the compound must be a "1,2,4-trisubstituted benzene". The proton at 7.48 ppm clearly has no *ortho* couplings and must be the isolated proton in the spin system sandwiched between two substituents. The two protons at 7.26 and 6.85 ppm each have a large *ortho* coupling so they must be adjacent to each other. The proton at 7.26 ppm has an additional *meta* coupling so it must be the proton *meta* to the proton at 7.48 ppm. The remaining proton at 6.85 ppm must be the proton *para* to the proton at 7.48 ppm.

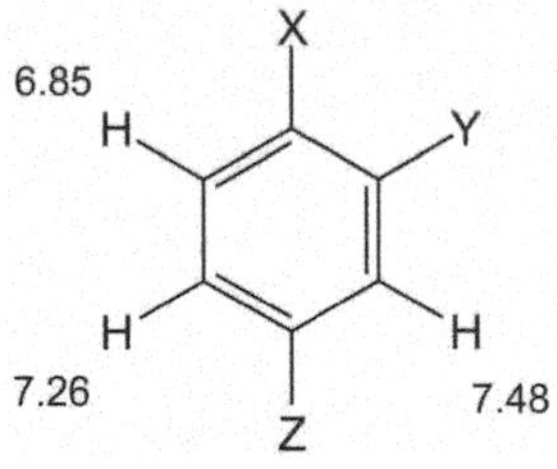

4. There are signals for two methyl groups at 2.59 and 2.30 ppm, one of which must be part of a methyl ketone (O=CCH$_3$). The remaining substituent on the aromatic ring is a hydroxy group (–OH, δ_H 12.08 ppm, exchangeable). It is not yet possible to establish which substituents are where.

^{1}H NMR spectrum of 2'-hydroxy-5'-methylacetophenone (CDCl$_3$, 600 MHz)

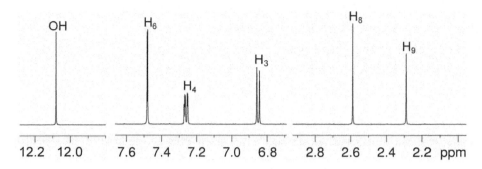

^{13}C{^{1}H} NMR spectrum of 2'-hydroxy-5'-methylacetophenone (CDCl$_3$, 150 MHz)

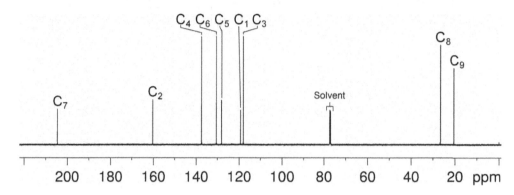

5. The ^{1}H–^{13}C me-HSQC spectrum easily identifies the protonated carbons. The isolated aromatic proton at 7.48 ppm correlates to the carbon at 130.6 ppm. The aromatic protons at 7.26 and 6.85 ppm correlate to the carbons at 137.5 and 118.1 ppm, respectively. The methyl protons at 2.59 and 2.30 ppm correlate to the carbons at 26.6 and 20.5 ppm, respectively.

6. The three aromatic carbons which bear the substituents appear at 160.3, 128.1 and 119.4 ppm. The low-field carbon resonance (160.3 ppm) must be the carbon bearing the –OH substituent based on its chemical shift.

^{1}H–^{13}C me-HSQC spectrum of 2'-hydroxy-5'-methylacetophenone (CDCl$_3$, 600 MHz)

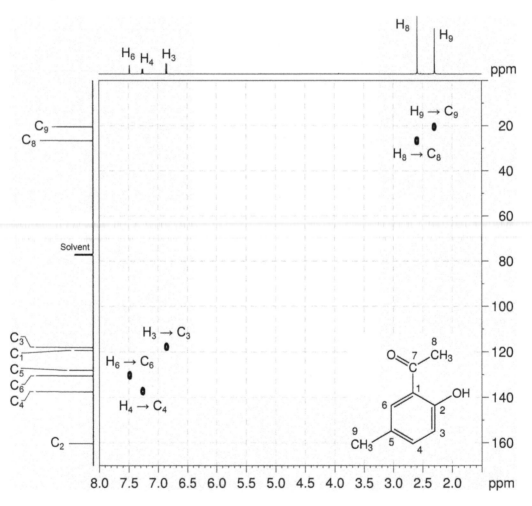

7.   For a "1,2,4-trisubstituted benzene", there are six possible isomers:

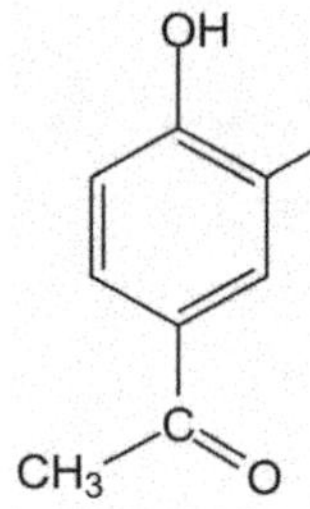

A

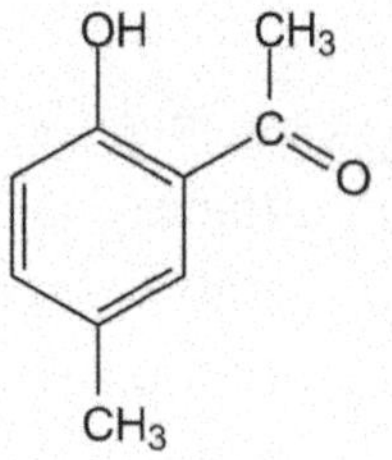

B

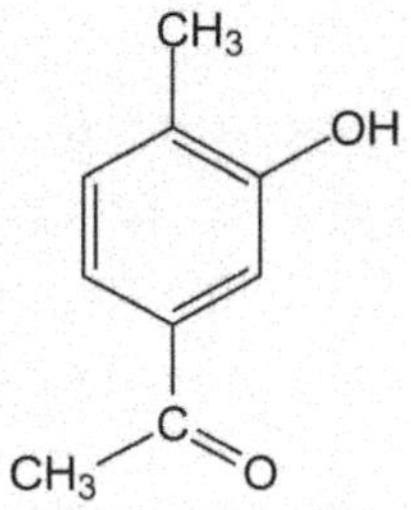

C

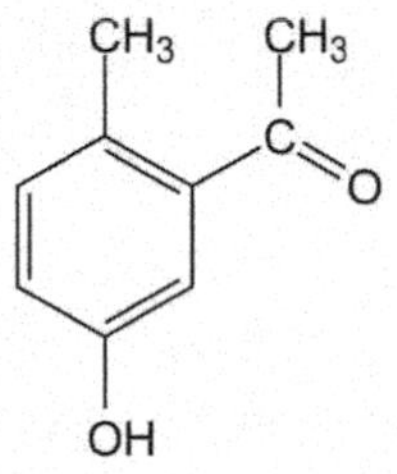

D

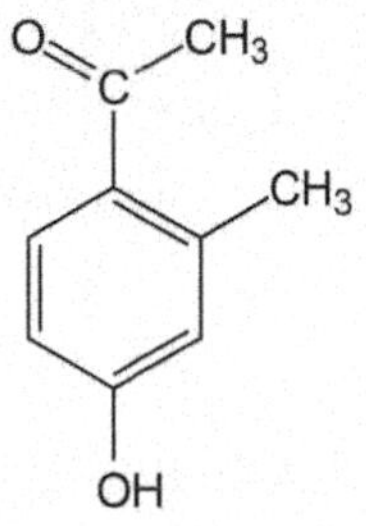

E

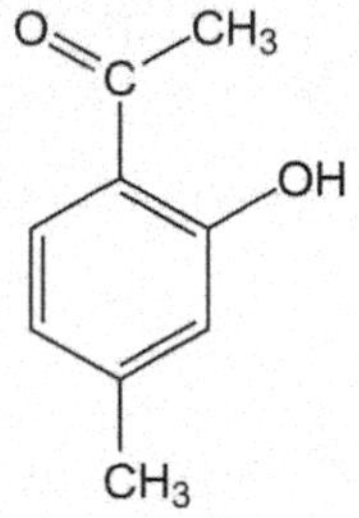

F

8. In the HMBC spectrum, the methyl proton resonance at 2.59 ppm correlates with the ketone resonance at 204.4 ppm identifying the methyl group attached to the ketone.

9. In the HMBC spectrum, the ^{1}H resonance for the other methyl group at 2.30 ppm correlates to protonated aromatic carbons at 130.6 and 137.5 ppm placing the methyl group between these two aromatic carbons. The reciprocal correlations from the aromatic protons at 7.48 and 7.26 ppm to methyl carbon at 20.5 ppm confirm the position of the methyl group. This eliminates Isomers A, C, D and E.

10. In the HMBC spectrum, the aromatic proton resonance at 7.48 ppm correlates to the ketone carbon at 204.4 ppm thus placing the methyl ketone group *ortho* to the proton at 7.48 ppm on the aromatic ring. This identifies **Isomer B** as the correct isomer.

^{1}H–^{13}C HMBC spectrum of 2'-hydroxy-5'-methylacetophenone (CDCl$_3$, 600 MHz)

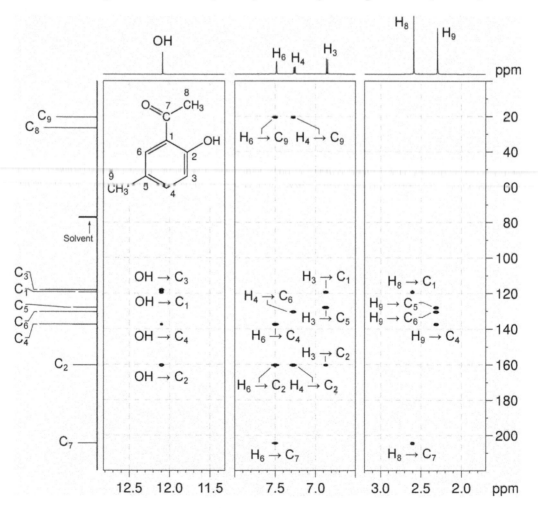

Question:

Identify the following compound.

Molecular Formula: $C_9H_9FO_2$

IR: 1681 cm^{-1}

Solution:

3'-Fluoro-4'-methoxyacetophenone

1. The molecular formula is $C_9H_9FO_2$. Calculate the degree of unsaturation from the molecular formula: replace the F by H and ignore the O atoms to give an effective molecular formula of C_9H_{10} (C_nH_m) which gives the degree of unsaturation as $(n - m/2 + 1) = 9 - 5 + 1 = 5$. The compound contains a combined total of five rings and/or π bonds.

2. The IR and $^{13}C\{^1H\}$ NMR spectra establish that the compound is aromatic and contains a ketone functional group (quaternary ^{13}C resonance at 195.9 ppm – although not visible in the spectra given, the ketone resonance is fluorine-coupled with $^4J_{C-F} = 1.8$ Hz). This accounts for all of the degrees of unsaturation, so the compound contains no additional rings or multiple bonds.

3. There are three aromatic proton resonances in the ^1H NMR spectrum. On the basis of the coupling pattern in the $^1H\{^{19}F\}$NMR spectrum, the compound must be a "1,2,4-trisubstituted benzene". The proton at 7.68 ppm clearly has no *ortho* couplings and must be the isolated proton in the spin system sandwiched between two substituents. The two protons at 7.75 and 7.20 ppm each have a large *ortho* coupling so they must be adjacent to each other. The proton at 7.75 ppm has an additional *meta* coupling so it must be the proton *meta* to the proton at 7.68 ppm. The remaining proton at 7.20 ppm must be the proton *para* to the proton at 7.68 ppm.

4. The strong fluorine coupling experienced by the aromatic protons suggests that the fluorine is bound directly to the aromatic ring.

5. The ¹H NMR spectrum has two three-proton resonances: the first at 3.90 ppm, and the second at 2.50 ppm. The chemical shift of the former is consistent with a methoxy group, –OCH₃, while the second is consistent with an acyl group (–C(=O)CH₃).

¹H NMR spectrum of 3'-fluoro-4'-methoxyacetophenone (DMSO-*d₆*, 400 MHz)

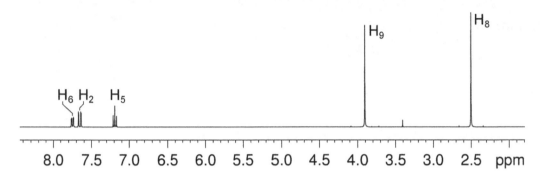

¹H and ¹H{¹⁹F} NMR spectra of 3'-fluoro-4'-methoxyacetophenone – expansions

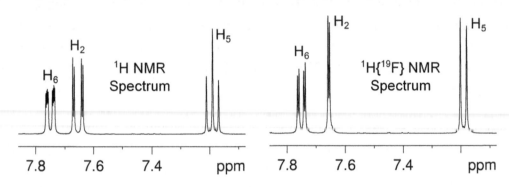

¹³C{¹H} NMR spectrum of 3'-fluoro-4'-methoxyacetophenone (DMSO-*d₆*, 100 MHz)

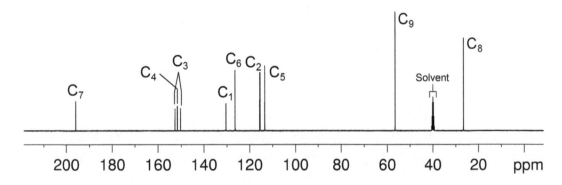

6. The ^{1}H–^{13}C me-HSQC spectrum easily identifies the
    protonated carbons. The isolated aromatic proton at 7.68
    ppm correlates to the carbon at 115.5 ppm. The aromatic
    protons at 7.75 and 7.20 ppm correlate to the carbons at 126.4
    and 113.4 ppm, respectively. The methyl protons at 3.90 and
    2.50 ppm correlate to the carbons at 56.5 and 26.6 ppm,
    respectively.

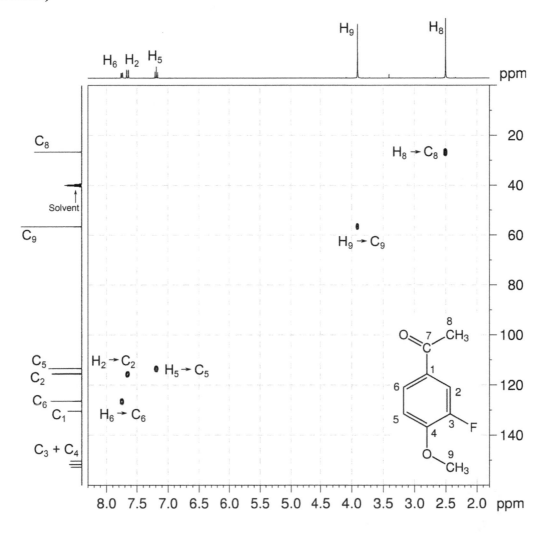

^{1}H–^{13}C me-HSQC spectrum of 3'-fluoro-4'-methoxyacetophenone (DMSO-d_6, 400 MHz)

7. For a "1,2,4-trisubstituted benzene", there are six possible isomers:

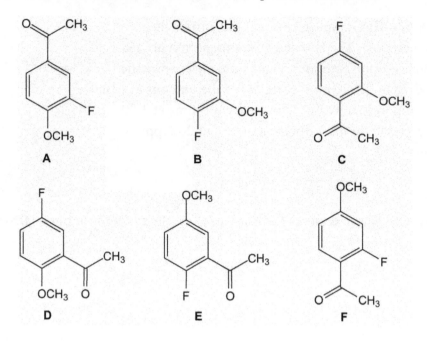

8. The ^{1}H–^{1}H NOESY spectrum identifies the relative substitution pattern. Remember that NOESY spectra are symmetrical about a diagonal, so we need only be concerned with cross-peaks *either* above *or* below the diagonal.

9. The acyl protons at 2.50 ppm are close in space to the aromatic protons at 7.75 and 7.68 ppm, placing the acyl group between these two protons on the aromatic ring. This eliminates isomers **C**, **D**, **E** and **F**.

10. In the ^{1}H–^{1}H NOESY spectrum, the aromatic proton at 7.20 ppm correlates to the methoxy group at 3.90 ppm thus placing the methoxy group *ortho* to the proton at 7.20 ppm on the aromatic ring. **Isomer A** (3'-fluoro-4'-methoxyacetophenone) is therefore the correct answer.

^{1}H–^{1}H **NOESY spectrum of 3'-fluoro-4'-methoxyacetophenone (DMSO-d_6, 400 MHz)**

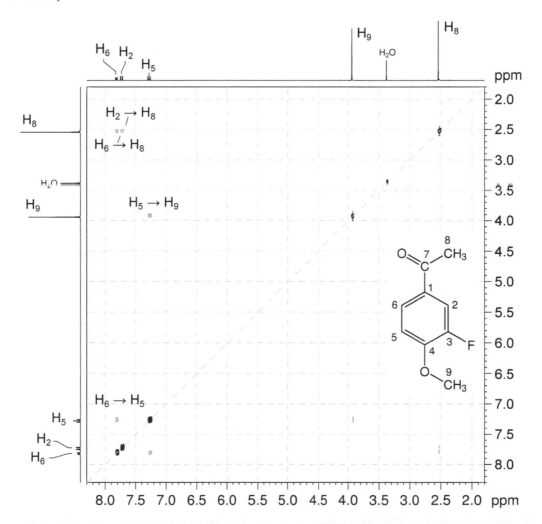

11. There are three quaternary aromatic carbon signals at 151.7, 151.5 and 130.3 ppm and these can be assigned by comparison of the $^{13}C\{^1H\}$ and $^{13}C\{^1H,^{19}F\}$ NMR spectra.

12. The ^{13}C signal at 151.5 ppm has the largest coupling to fluorine ($^1J_{C-F} = 245$ Hz) and must be due to the aromatic carbon directly bound to the fluorine atom (C_3).

13. The ^{13}C signal at 151.7 ppm has a larger coupling to fluorine ($^2J_{C-F} = 10$ Hz) than the signal at 130.3 ppm ($^3J_{C-F} = 5$ Hz) and can be assigned to C_4. The signal at 130.3 ppm can thus be assigned to C_1.

$^{13}C\{^1H\}$ NMR spectrum of 3'-fluoro-4'-methoxyacetophenone – expansions

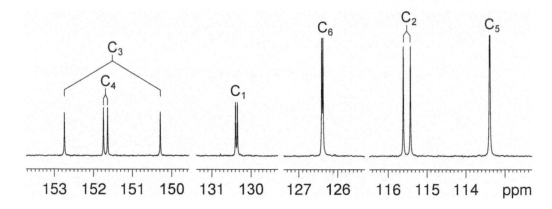

$^{13}C\{^1H,^{19}F\}$ NMR spectrum of 3'-fluoro-4'-methoxyacetophenone – expansions

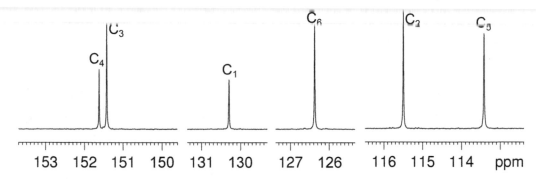

Problem 43

Question:

Identify the following compound.

Molecular Formula: $C_{10}H_{10}O_4$

Solution:

trans-Ferulic acid

1. The molecular formula is $C_{10}H_{10}O_4$. Calculate the degree of unsaturation from the molecular formula: ignore the O atoms to give an effective molecular formula of $C_{10}H_{10}$ (C_nH_m) which gives the degree of unsaturation as $(n - m/2 + 1) = 10 - 5 + 1 = 6$. The compound contains a combined total of six rings and/or π bonds.

2. 1D NMR data establish that the compound is aromatic and contains an alkene (¹H signals at 7.65, 6.42 ppm, both doublets with $^3J_{H-H} = 16$ Hz thus alkene protons are *trans* to each other) and a carbonyl functional group (¹³C signal at 167.7 ppm). This accounts for all of the degrees of unsaturation, so the compound contains no additional rings or multiple bonds.

3. 1D NMR data also establish the presence of two hydroxyl groups (¹H signals at 10.5 and 8.15 ppm, exchangeable) and a methoxy group (¹H signal at 3.95 ppm, ¹³C signal at 55.5 ppm).

4. The coupling pattern in the expansion of the aromatic region of the ¹H NMR spectrum shows that the compound must be a "1,2,4-trisubstituted benzene". The proton at 7.36 ppm clearly has no *ortho* couplings and must be the isolated proton sandwiched between two substituents. The two protons at 7.17 and 6.91 ppm each have a large *ortho* coupling so they must be adjacent to each other. The proton at 7.17 ppm has an additional *meta* coupling so it must be the proton *meta* to the proton at 7.36 ppm. The remaining proton at 6.91 ppm must be the proton *para* to the proton at 7.36 ppm.

^{1}H NMR spectrum of *trans*-ferulic acid (acetone-d_6, 600 MHz)

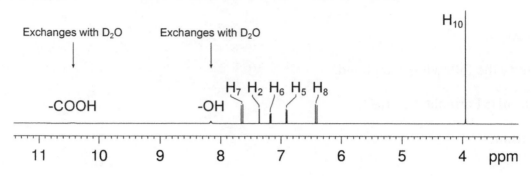

^{1}H NMR spectrum of *trans*-ferulic acid – expansion

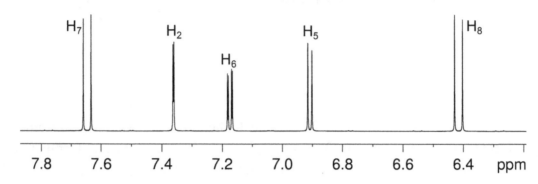

^{13}C{^{1}H} NMR spectrum of *trans*-ferulic acid (acetone-d_6, 150 MHz)

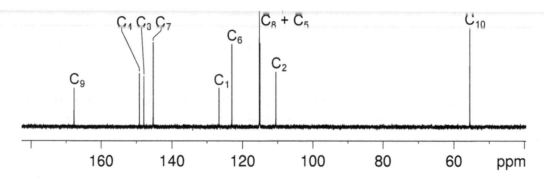

5. The COSY spectrum shows the expected correlations between the alkene protons at 7.65 and 6.42 ppm. It also shows the expected correlations between the aromatic protons.

^{1}H–^{1}H COSY spectrum of *trans*-ferulic acid (acetone-d_6, 600 MHz)

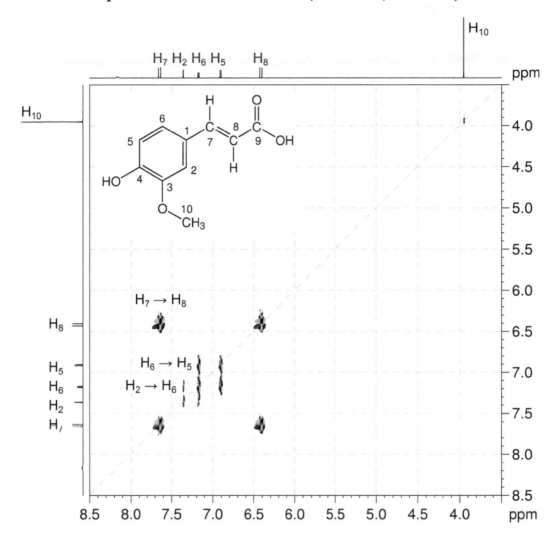

6. The $^1H-^{13}C$ me-HSQC spectrum easily identifies the protonated carbons. The alkene protons at 7.65 and 6.42 ppm correlate to carbons at 145.2 and 115.1 ppm, respectively. The aromatic proton at 7.36 ppm correlates to the carbon at 110.5 ppm. The aromatic protons at 7.17 and 6.91 ppm correlate to carbons at 123.0 and 115.3 ppm, respectively.

7. The three aromatic carbons which bear substituents appear at 149.2, 147.9 and 126.6 ppm. The low-field resonances at 149.2 and 147.9 ppm are consistent with aromatic carbons bound to oxygen.

$^1H-^{13}C$ me-HSQC spectrum of *trans*-ferulic acid (acetone-d_6, 600 MHz)

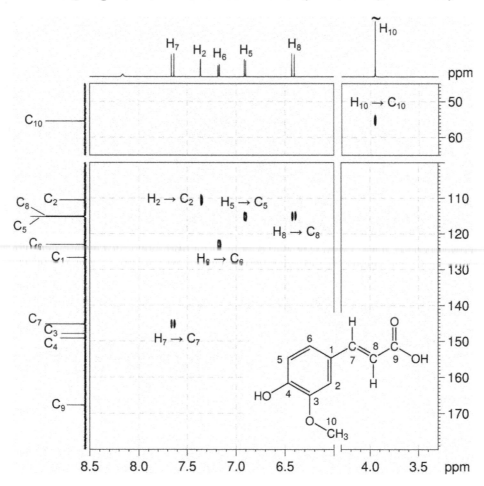

8. In the HMBC spectrum, the alkene proton at 7.65 ppm correlates to the protonated aromatic carbons at 110.5 and 123.0 ppm, thus placing the alkene group between these two aromatic carbons (*i.e.* at position Z on the ring). The reciprocal correlations from the aromatic protons at 7.36 and 7.17 ppm to the alkene carbon at 145.2 ppm confirm the position of the alkene group.

9. Remember that, in aromatic systems, the three-bond coupling $^3J_{C-H}$ is typically the larger long-range coupling and gives rise to the strongest cross-peaks.

10. In the HMBC spectrum, the correlations from the aromatic protons at 7.36 and 7.17 ppm to the quaternary aromatic carbon signal at 149.2 ppm identify the carbon *meta* to both these protons (*i.e.* the carbon bearing substituent X). The correlations from the aromatic proton at 6.91 and the alkene proton at 6.42 ppm to the quaternary aromatic carbon signal at 126.6 ppm identify the carbon three-bonds away from both these protons (*i.e.* the carbon bearing substituent Z). The correlation from the aromatic proton at 6.91 ppm to the quaternary aromatic carbon signal at 147.9 ppm identifies the carbon *meta* to this proton (*i.e.* the carbon bearing substituent Y).

11. In the HMBC spectrum, the correlation of the methoxy group at 3.95 ppm to the carbon at 147.9 ppm places the methoxy group in position Y of the ring.

12. The correlation from the alkene proton at 7.65 ppm to the carbonyl carbon at 167.7 ppm places the carbonyl group next to the alkene functionality.

13. By elimination, one hydroxyl group must be in position X of the ring and the other bound to the carbonyl group to give a carboxylic acid.

^{1}H–^{13}C HMBC spectrum of *trans*-ferulic acid – expansion (acetone-d_6, 600 MHz)

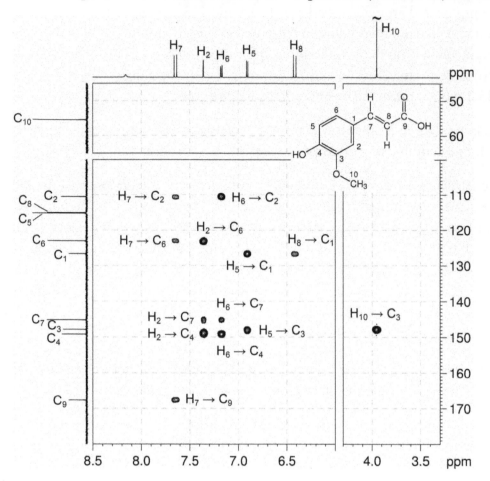

Problem 44

Question:

Identify the following compound.

Molecular Formula: $C_{13}H_{16}O_3$

IR: 3385 (br), 2973, 1685, 1636, 1583 cm^{-1}

Solution:

sec-Butyl 3-hydroxycinnamate

1. The molecular formula is $C_{13}H_{16}O_3$. Calculate the degree of unsaturation from the molecular formula: ignore the O atoms to give an effective molecular formula of $C_{13}H_{16}$ (C_nH_m) which gives the degree of unsaturation as $(n - m/2 + 1) = 13 - 8 + 1 = 6$. The compound contains a combined total of six rings and/or π bonds.

2. IR and 1D NMR data establish the presence of an alkene (^{1}H resonances at 7.61 and 6.46 ppm, both doublets with $^3J_{H-H} = 16$ Hz thus alkene protons are *trans* to each other, II_7 and II_8, respectively), a disubstituted benzene ring (^{1}H resonances at 7.26, 7.14, 7.12 and 6.92 ppm) and an ester group (^{13}C resonance at 165.9 ppm). This accounts for all of the degrees of unsaturation, so the compound contains no additional rings or multiple bonds.

3. 1D NMR data also establish the presence of a hydroxyl (^{1}H resonance at 8.58 ppm, exchangeable), an OCH group (^{1}H resonance at 4.93 ppm, H_{10}), a CH_2 group (^{1}H resonance at 1.63 ppm, H_{11}) and two methyl groups (^{1}H resonances at 1.25 and 0.93 ppm, H_{13} and H_{12}, respectively). The multiplicities of the signals are confirmed by the ^{1}H–^{13}C me-HSQC spectrum.

4. The coupling pattern in the aromatic region of the ^{1}H NMR spectrum establishes that the benzene ring is 1,3-disubstituted with the signal at 7.12 ppm due to the proton between the two substituents (H_2) and the signal at 7.26 ppm due to the proton between two aromatic protons (H_5).

^{1}H NMR spectrum of *sec*-butyl 3-hydroxycinnamate (acetone-d_6, 500 MHz)

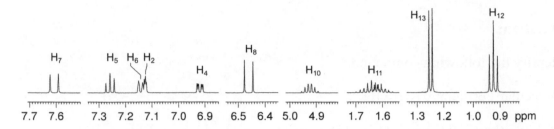

^{13}C{^{1}H} NMR spectrum of *sec*-butyl 3-hydroxycinnamate (acetone-d_6, 125 MHz)

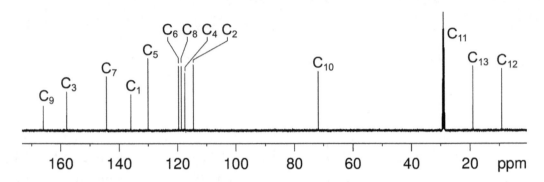

5. The COSY spectrum shows the expected correlations between the two alkene protons (H_7–H_8) and aromatic protons.

6. The COSY spectrum also shows correlations from methyl protons H_{12} to methylene protons H_{11}, then from methylene protons H_{11} to protons of OCH group H_{10}, and lastly from OCH protons H_{10} to methyl protons H_{13} which afford the $OCH(CH_3)CH_2CH_3$ fragment.

^1H–^1H COSY spectrum of *sec*-butyl 3-hydroxycinnamate (acetone-d_6, 500 MHz)

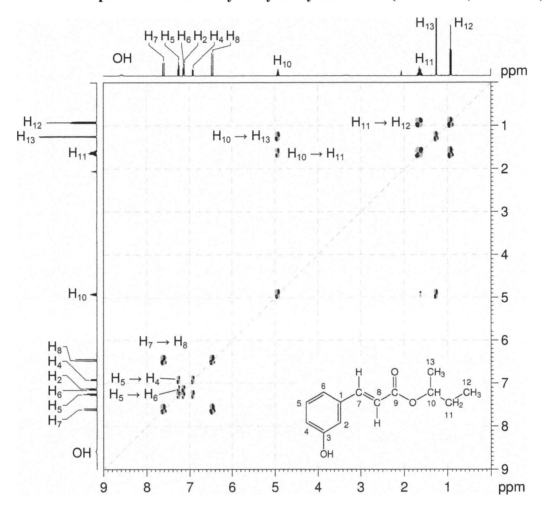

7. The protonated carbons are easily identified by the me-HSQC spectrum: C_{10}, C_{11}, C_{13} and C_{12} at 71.8, 28.7, 19.0 and 9.2 ppm, respectively. From expansion A: C_7, C_5, C_6, C_8, C_4 and C_2 at 144.2, 130.0, 119.6, 118.7, 117.5 and 114.5 ppm, respectively.

¹H–¹³C me-HSQC spectrum of *sec*-butyl 3-hydroxycinnamate (acetone-d_6, 500 MHz)

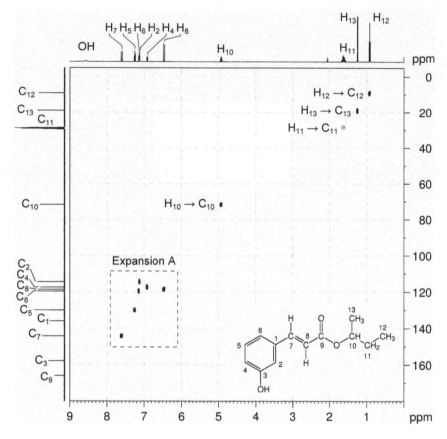

¹H–¹³C me-HSQC spectrum of *sec*-butyl 3-hydroxycinnamate – expansion A

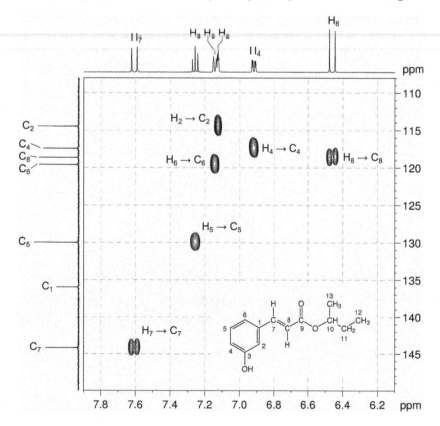

8. In expansion B of the HMBC spectrum, all correlations are consistent with the $OCH(CH_3)CH_2CH_3$ fragment deduced earlier using the COSY spectrum.

9. Note that the one bond couplings between H_{12} and C_{12} and between H_{13} and C_{13} are visible as large doublets.

^1H–^{13}C HMBC spectrum of *sec*-butyl 3-hydroxycinnamate – expansion B (acetone-d_6, 500 MHz)

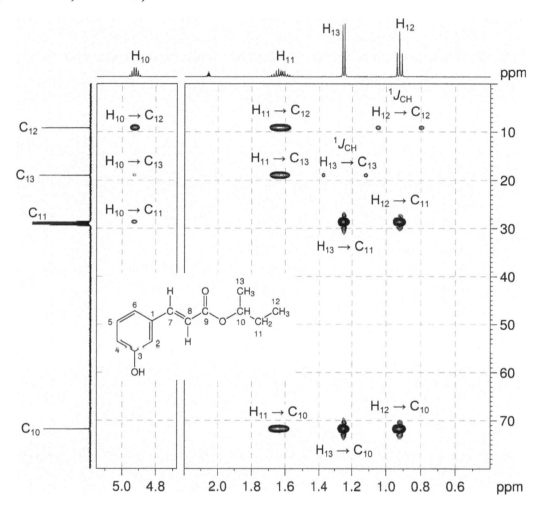

10. In expansion C of the HMBC spectrum, the alkene to carbonyl correlations (H_7–C_9, H_8–C_9) and the OCH to carbonyl correlation (H_{10}–C_9) places the carbonyl group between the alkene and OCH groups.

11. In the HMBC spectrum, the alkene proton H_7 to protonated aromatic carbon C_2 / C_6 correlations as well as the aromatic proton H_2 / H_6 to alkene carbon C_7 correlations place the alkene group in position 1 of the ring.

14. By elimination, the hydroxyl group must be located at position 3 on the benzene ring. The downfield chemical shift of C_3 is consistent with a quaternary aromatic carbon bound to oxygen.

^{1}H–^{13}C HMBC spectrum of *sec*-butyl 3-hydroxycinnamate – expansion C

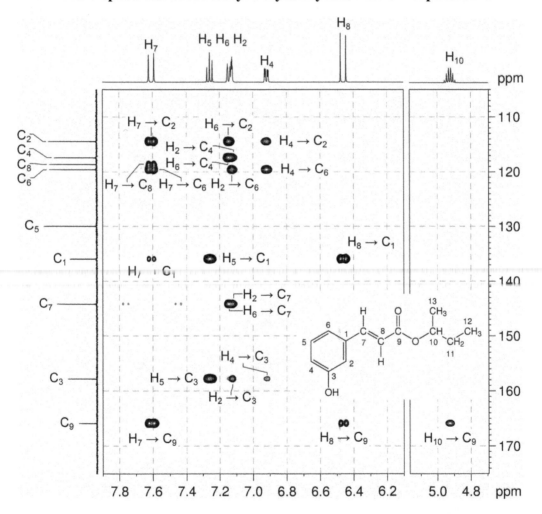

15. The H_7, H_5 and H_8 correlations to the quaternary aromatic carbon signal at 136.0 ppm identifies the resonance as due to C_1.

16. The H_5, H_2 and H_4 correlations to the quaternary aromatic carbon signal at 157.9 ppm identifies the resonance as due to C_3.

Question:

Identify the following compound. Draw a labeled structure, and use the $^1H-^1H$ COSY, $^1H-^{13}C$ me-HSQC, and $^1H-^{13}C$ HMBC spectra to assign each of the ^1H and ^{13}C resonances to the appropriate carbons and hydrogens in the structure.

Molecular Formula: $C_{11}H_{12}O$

IR: 1676 cm^{-1}

Solution:

1-Benzosuberone

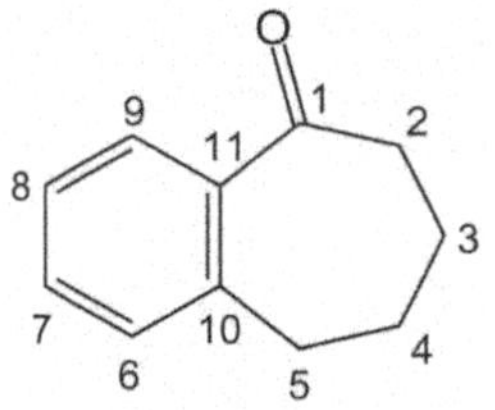

Proton	Chemical Shift (ppm)	Carbon	Chemical Shift (ppm)
		C_1	207.0
H_2	2.70	C_2	40.8
H_3	1.77	C_3	20.9
H_4	1.85	C_4	25.2
H_5	2.89	C_5	32.4
H_6	7.17	C_6	129.7
H_7	7.38	C_7	132.1
H_8	7.26	C_8	126.5
H_9	7.70	C_9	128.5
		C_{10}	141.3
		C_{11}	138.8

1. The molecular formula is $C_{11}H_{12}O$. Calculate the degree of unsaturation from the molecular formula: ignore the O atom to give an effective molecular formula of $C_{11}H_{12}$ (C_nH_m) which gives the degree of unsaturation as $(n - m/2 + 1) = 11 - 6 + 1 = 6$. The compound contains a combined total of six rings and/or π bonds.

2. The coupling pattern in the aromatic region of the ^{1}H NMR spectrum identifies an *ortho*-disubstituted aromatic ring. From the integration of the alkyl region of the spectrum, it is clear that there are four CH$_2$ groups, all of which show complicated coupling patterns.

3. The ^{13}C$\{^1$H$\}$ NMR spectrum shows a signal to low-field (207.0 ppm), consistent with the presence of a ketone functional group.

4. The aromatic ring and ketone functional group accounts for five degrees of unsaturation and the remaining degree of unsaturation may be due to a ring.

^{1}H NMR spectrum of 1-benzosuberone (CDCl$_3$, 500 MHz)

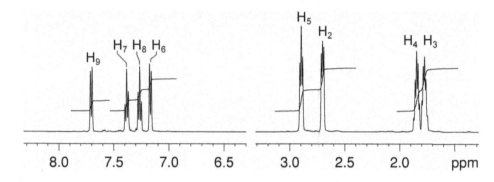

^{13}C$\{^1$H$\}$ NMR spectrum of 1-benzosuberone (CDCl$_3$, 125 MHz)

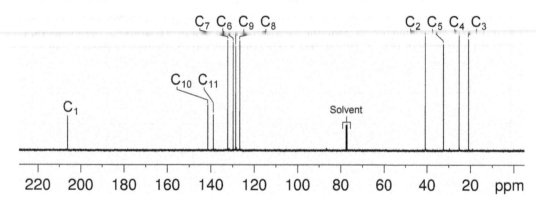

5. The ^1H–^1H COSY spectrum (expansion A) indicates that all the CH_2 groups are sequentially linked – affording a $CH_2CH_2CH_2CH_2$ chain. Expansion B confirms that the aromatic ring is 1,2-disubstituted.

^1H–^1H COSY spectrum of 1-benzosuberone – expansion A (CDCl₃, 500 MHz)

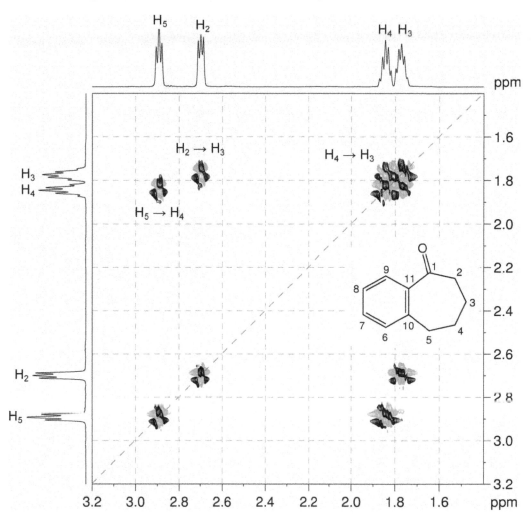

^1H–^1H COSY spectrum of 1-benzosuberone – expansion B

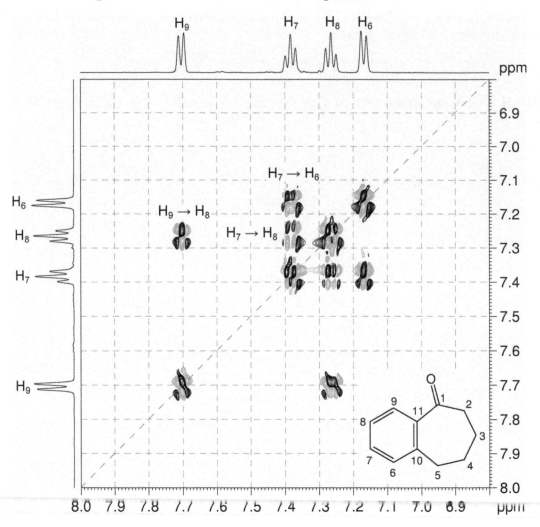

6. The molecular formula is now accounted for, and the known fragments are:

7. There is only one way to assemble these fragments, so the compound must be 1-benzosuberone:

8. We can now begin to assign each resonance to its corresponding nucleus. The carbonyl carbon (C_1) is the most readily identified, so we will build out from there, initially using the HMBC spectrum.

^{1}H–^{13}C HMBC spectrum of 1-benzosuberone (CDCl₃, 500 MHz)

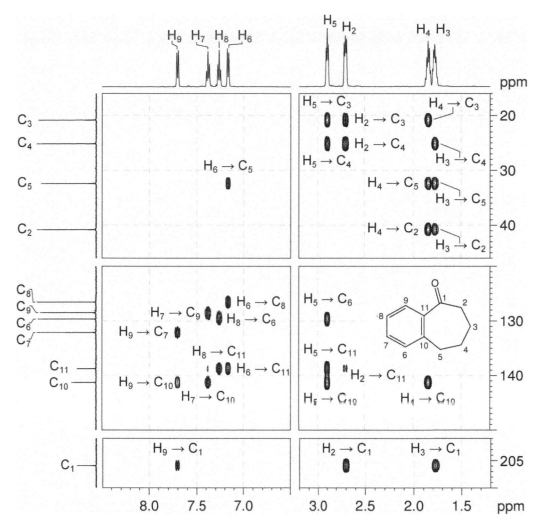

9. There is one correlation from an aromatic proton to the carbonyl carbon. Remember that, in aromatic systems, the three-bond coupling $^3J_{\text{C–H}}$ is typically the larger long-range coupling and gives rise to the strongest cross-peaks. Three-bond couplings to benzylic carbon atoms are also large. Using this information, it is possible to assign the low-field aromatic resonance at 7.70 ppm to H_9 – as this proton is three bonds removed from the carbonyl carbon atom.

10. Using the ^{1}H–^{13}C me-HSQC spectrum, we can identify C_9 as the resonance at 128.5 ppm.

^{1}H–^{13}C me-HSQC spectrum of 1-benzosuberone (CDCl$_3$, 500 MHz)

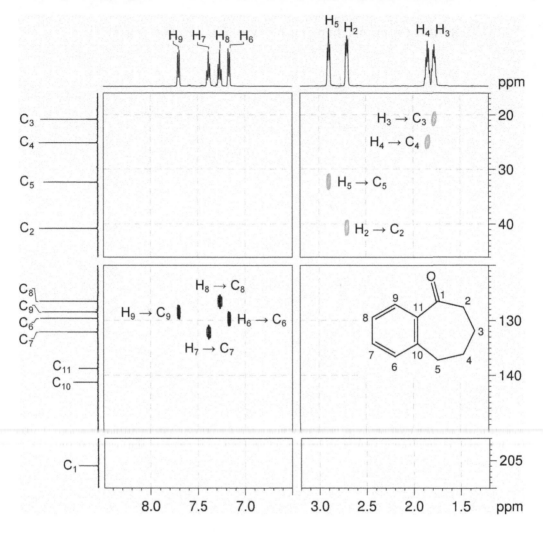

11. In the HMBC spectrum, H_9 should also correlate to the two aromatic ring carbon atoms *meta* to it, *i.e.* C_7 and C_{10}. Indeed, H_9 correlates to the carbon resonances at 132.1 and 141.3 ppm.

12. Using the ^{1}H–^{13}C me-HSQC spectrum, we can demonstrate that the resonance at 132.1 ppm corresponds to a protonated carbon nucleus, and therefore must be C_7. H_7 is identified as the signal at 7.38 ppm. C_{10} is the resonance at 141.3 ppm.

13. As the signal at 138.8 ppm is also a quaternary carbon atom, this must correspond to C_{11}.

14. Using the COSY spectrum (or the coupling patterns in the 1D proton spectrum) we can identify H_8 as the signal at 7.26 ppm, and H_6 as the signal at 7.17 ppm. From the me-HSQC spectrum, C_8 and C_6 are assigned to the signals at 126.5 and 129.7 ppm, respectively.

15. In the HMBC spectrum, the alkyl signals at 2.70 and 1.77 ppm both correlate to the carbonyl carbon, but from the ^{1}H–^{1}H COSY spectrum, we know that the signal at 2.70 ppm is due to a terminal CH_2, and must therefore be H_2. The ^{1}H resonance at 1.77 ppm is assigned to H_3.

16. Using the ^{1}H–^{13}C me-HSQC spectrum, C_2 is identified as the resonance at 40.8 ppm and C_3 as the resonance at 20.9 ppm.

17. From the ^{1}H–^{1}H COSY spectrum, we can identify H_4 (which couples to H_3) as the resonance at 1.85 ppm and H_5 (which couples to H_4) as the resonance at 2.89 ppm.

18. Using the ^{1}H–^{13}C me-HSQC spectrum, C_4 is identified as the resonance at 25.2 ppm and C_5 as the resonance at 32.4 ppm.

Problem 46

Question:

Identify the following compound. Draw a labeled structure, and use the ^1H–^{13}C me-HSQC, ^1H–^1H COSY and ^1H–^{31}P HMBC spectra to assign each of the ^1H and ^{13}C resonances to the appropriate carbons and hydrogens in the structure.

Molecular Formula: $C_5H_{12}BrO_3P$

IR: 3464, 2955, 1232, 1031 cm^{-1}

Solution:

Dimethyl (3-bromopropyl)phosphonate

$$CH_3O\diagdown$$
$$P-\underset{2}{CH_2}-\underset{3}{CH_2}-\underset{4}{CH_2}-Br$$
$$CH_3O\diagup \underset{1}{\Vert} \; O$$

Proton	Chemical Shift (ppm)	Carbon	Chemical Shift (ppm)
H_1	3.69	C_1	52.5
H_2	1.87	C_2	23.3
H_3	2.09	C_3	25.8
H_4	3.42	C_4	33.5

1. The molecular formula is $C_5H_{12}BrO_3P$. The method for calculating the degree of unsaturation does not work for this compound as it contains a P atom.

2. ID spectra establish that there are two methoxy (OCH_3) groups (^1H resonance at 3.69 ppm, ^{13}C resonance at 52.5 ppm) and three CH_2 groups (^1H resonances at 3.42, 2.09 and 1.87 ppm). This accounts for all the C and H atoms as well as two of the three O atoms in the molecule.

^{1}H NMR spectrum of dimethyl (3-bromopropyl)phosphonate (CDCl$_3$, 400 MHz)

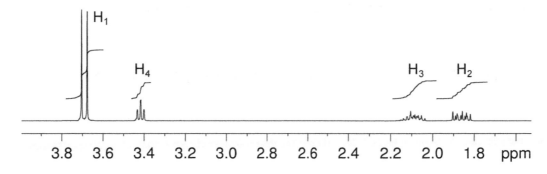

^{1}H{^{31}P} NMR spectrum of dimethyl (3-bromopropyl)phosphonate (CDCl$_3$, 400 MHz)

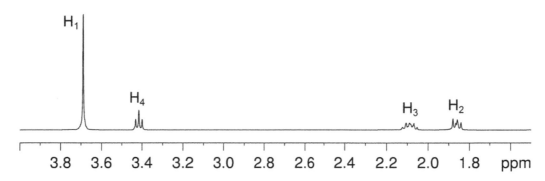

^{13}C{^{1}H} NMR spectrum of dimethyl (3-bromopropyl)phosphonate (CDCl$_3$, 100 MHz)

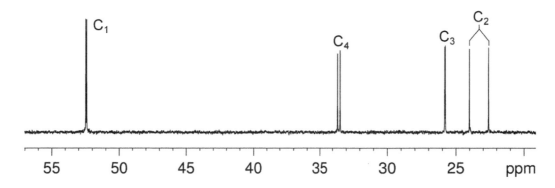

3.   In the COSY spectrum, the H$_2$–H$_3$ and H$_3$–H$_4$ correlations indicate that the CH$_2$ protons are sequentially linked affording a –CH$_2$CH$_2$CH$_2$– fragment.

^{1}H–^{1}H COSY spectrum of dimethyl (3-bromopropyl)phosphonate (CDCl$_3$, 400 MHz)

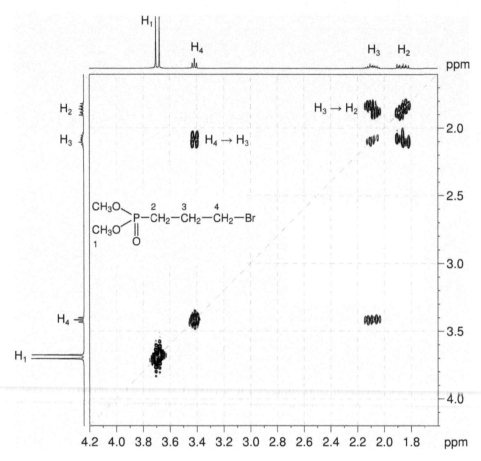

4. In the me-HSQC spectrum, the ^{1}H signal at 3.42 ppm correlates with the ^{13}C signal at 33.5 ppm. The ^{13}C chemical shift is consistent with those for a CH_2–Br group rather than a CH_2O group (^{13}C shifts for CH_2O groups are in the 50–75 ppm range). This affords a $-CH_2CH_2CH_2Br$ fragment.

5. Note that the C_2 resonance on the vertical axis is split into a large doublet by a coupling to ^{31}P and that the H_2–C_2 correlation is similarly split.

^{1}H–^{13}C me-HSQC spectrum of dimethyl (3-bromopropyl)phosphonate (CDCl$_3$, 400 MHz)

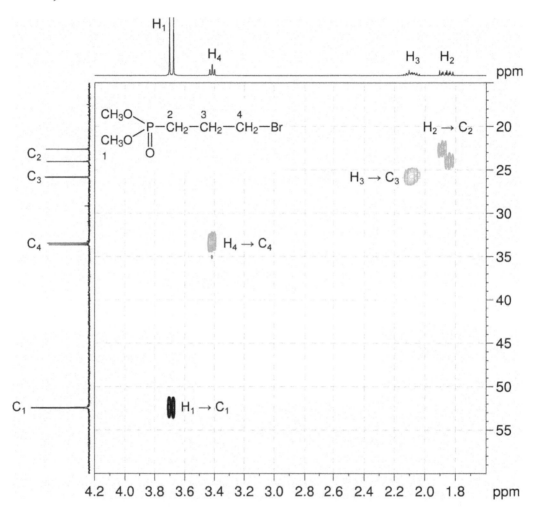

6. The ^{1}H–^{31}P HMBC spectrum shows two- and three-bond correlations between proton and phosphorus nuclei. In the ^{1}H–^{31}P HMBC spectrum, there are correlations from methoxy protons H_1 to the phosphorus signal, as well as from methylene protons H_2 and H_3 to the phosphorus signal. The methoxy groups must be bound to phosphorus to afford the $(CH_3O)_2P$ fragment.

7. The bromopropyl group must also be bound to phosphorus as there are correlations from H_2 and H_3 to phosphorus in the ^{1}H–^{31}P HMBC spectrum.

^{1}H–^{31}P HMBC spectrum of dimethyl (3-bromopropyl)phosphonate (CDCl₃, 400 MHz)

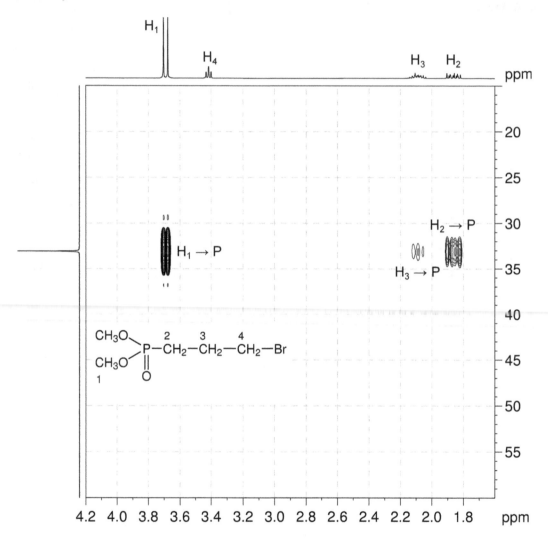

8. The remaining oxygen atom must be bound to P with a double bond to satisfy the valency of the P atom.

9. The ^{13}C resonances can now be assigned using the ^{1}H–^{13}C me-HSQC spectrum.

10. Note that there is a large $^1J_{C–P}$ coupling of 143 Hz for C_2.

Question:

The ^1H and $^{13}C\{^1H\}$ NMR spectra of caffeine ($C_8H_{10}N_4O_2$) recorded in DMSO-d_6 solution at 298 K and 500 MHz are given below.

The ^1H NMR spectrum has signals at δ 3.16, 3.35, 3.84 and 7.96 ppm.

The $^{13}C\{^1H\}$ NMR spectrum has signals at δ 27.4, 29.3, 33.1, 106.5, 142.7, 148.0, 150.9 and 154.4 ppm.

The ^{15}N NMR spectrum has signals at δ 112.9, 149.6, 156.1 and 231.2 ppm.

The multiplicity-edited ^1H–^{13}C HSQC, ^1H–^{13}C HMBC and ^1H–^{15}N HMBC spectra are given on the following pages. Use these spectra to assign each ^1H, ^{13}C and ^{15}N resonance to its corresponding nucleus.

Solution:

Caffeine

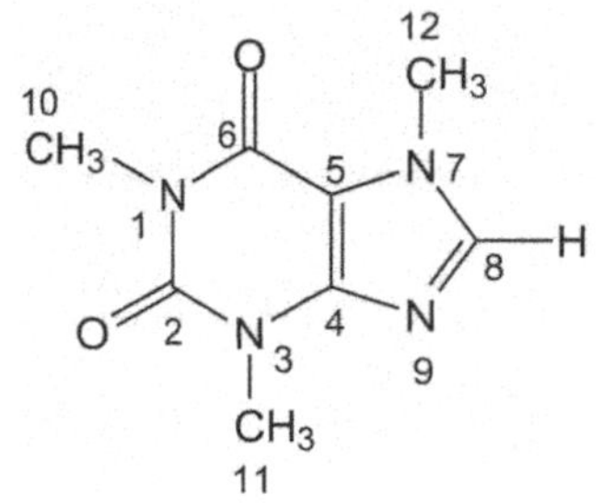

Proton	Chemical Shift (ppm)	Nucleus	Chemical Shift (ppm)
		N_1	149.6
		C_2	150.9
		N_3	112.9
		C_4	148.0
		C_5	106.5
		C_6	154.4
		N_7	156.1
H_8	7.96	C_8	142.7
		N_9	231.2
H_{10}	3.16	C_{10}	27.4
H_{11}	3.35	C_{11}	29.3
H_{12}	3.84	C_{12}	33.1

^{1}H NMR spectrum of caffeine (DMSO-d_6, 500 MHz)

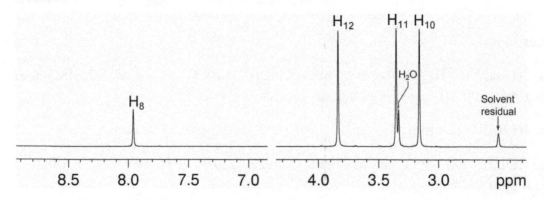

^{13}C{^{1}H} NMR spectrum of caffeine (DMSO-d_6, 500 MHz)

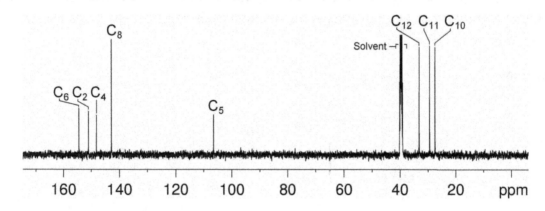

1. The most readily identifiable signal is that corresponding to H_8, at 7.96 ppm. From the $^1H–^{13}C$ me-HSQC spectrum we can identify C_8 as the signal at 142.7 ppm.

$^1H–^{13}C$ me-HSQC spectrum of caffeine (DMSO-d_6, 500 MHz)

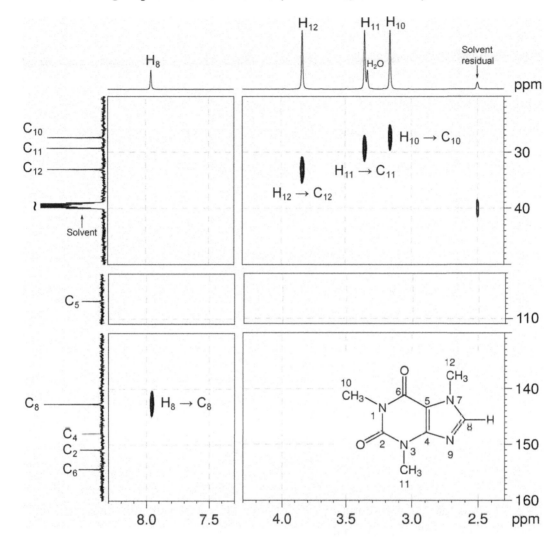

2. In the $^1H–^{13}C$ HMBC spectrum, there is a correlation between the proton resonance at 3.84 ppm and C_8. H_{10} and H_{11} are too far removed from C_8 to correlate, so this signal must be due to H_{12}.

3. Using the HSQC spectrum, we can then identify C_{12} as the signal at 33.1 ppm.

^{1}H–^{13}C HMBC spectrum of caffeine (DMSO-d_6, 500 MHz)

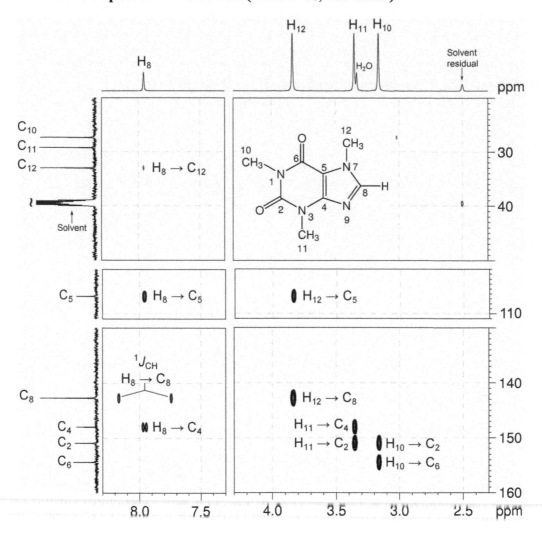

4. In the ^{1}H–^{13}C HMBC spectrum, there is also a correlation between the resonance H$_{12}$ (3.84 ppm) and the ^{13}C resonance at 106.5 ppm. This signal must be assigned to C$_5$, as H$_{12}$ is three bonds removed from this carbon nucleus.

5. In the ^{1}H–^{13}C HMBC spectrum, H$_8$ shows three correlations – a very weak correlation to C$_{12}$, a correlation to C$_5$, and a correlation to the signal at 148.0 ppm. This signal must be due to C$_4$.

6. The methyl signal at 3.35 ppm also correlates to C$_4$. As H$_{10}$ is too far removed to correlate to this position, this resonance must be due to H$_{11}$.

7. C$_{11}$ is identified from the me-HSQC spectrum as the resonance at 29.3 ppm.

8. In the ^{1}H–^{13}C HMBC spectrum, H$_{11}$ also correlates to the resonance at 150.9 ppm. The resonance at 150.9 ppm is therefore assigned to C$_2$.

9. By elimination, the remaining signal in the ^{1}H NMR spectrum (at 3.16 ppm) is due

to H_{10}. This resonance correlates to the signal at 27.4 ppm in the me-HSQC spectrum (C_{10}) and to the signals at 150.9 (C_2) and 154.4 ppm in the $^1H-^{13}C$ HMBC spectrum. The latter signal is therefore identified as C_6.

10. There is an unsuppressed one-bond correlation in the $^1H-^{13}C$ HMBC spectrum (between H_8 and C_8).

11. The $^1H-^{15}N$ HMBC spectrum shows two- and three-bond correlations between proton and nitrogen nuclei. The chemical shifts of the nitrogen nuclei can be determined using the $^1H-^{15}N$ HMBC spectrum:

   a. H_{10} correlates to the signal at 149.6 ppm, which must be N_1.

   b. H_{11} correlates to the signal at 112.9 ppm, which must be N_3.

   c. H_{12} correlates to the signal at 156.1 ppm, which must be N_7.

   d. H_8 correlates to the signals at 156.1 ppm (N_7) and 231.2 ppm. The latter signal must correspond to N_9.

$^1H-^{15}N$ HMBC spectrum of caffeine (DMSO-d_6, 500 MHz)

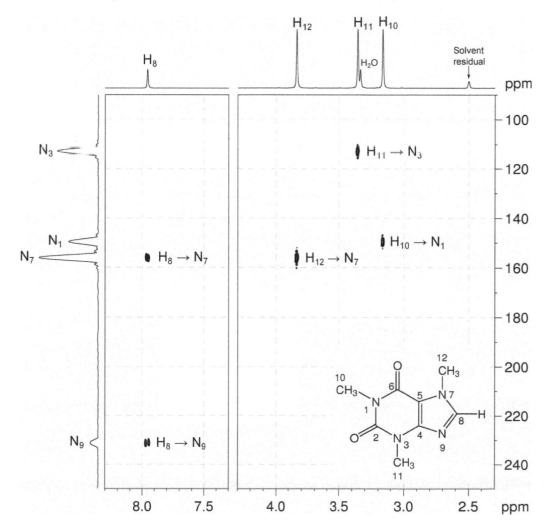

Problem 48

Question:

Identify the following compound.

Molecular Formula: $C_{10}H_{11}NO$

IR: 2252 cm^{-1}

Solution:

Benzyloxypropionitrile

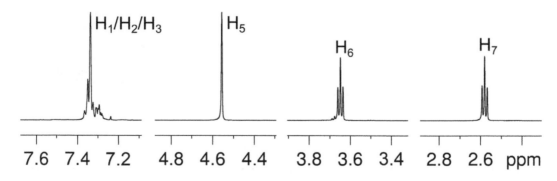

1. The molecular formula is $C_{10}H_{11}NO$. Calculate the degree of unsaturation from the molecular formula: ignore the O atom and ignore the N and remove one H to give an effective molecular formula of $C_{10}H_{10}$ (C_nH_m) which gives the degree of unsaturation as $(n - m/2 + 1) = 10 - 5 + 1 = 6$. The compound contains a combined total of six rings and/or π bonds.

2. 1D NMR data establish that the compound is aromatic which accounts for four degrees of unsaturation. There are two remaining degrees of unsaturation.

3. IR (2252 cm^{-1}) and ^{13}C NMR (resonance at 117.9 ppm) spectra indicate the possibility of a nitrile group.

4. In the ^1H NMR spectrum, there are overlapping signals for a phenyl ring at 7.40–7.25 ppm (total integration 5H, $H_1/H_2/H_3$), two OCH$_2$ groups at 4.56 and 3.65 ppm (H_5 and H_6, respectively) and one CH$_2$ group at 2.58 ppm (H_7). The multiplicities of the signals can be verified by the me-HSQC spectrum.

^1H NMR spectrum of benzyloxypropionitrile (CDCl$_3$, 500 MHz)

^{13}C{^{1}H} NMR spectrum of benzyloxypropionitrile (CDCl$_3$, 125 MHz)

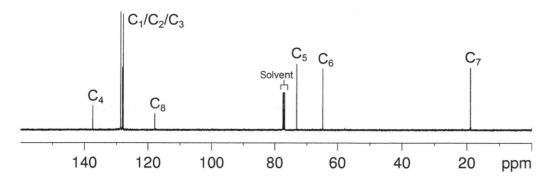

5. Both ^{1}H NMR and COSY spectra show that the H$_6$ and H$_7$ protons are coupled to afford a OCH$_2$CH$_2$ fragment.

^{1}H–^{1}H COSY spectrum of benzyloxypropionitrile (CDCl$_3$, 500 MHz)

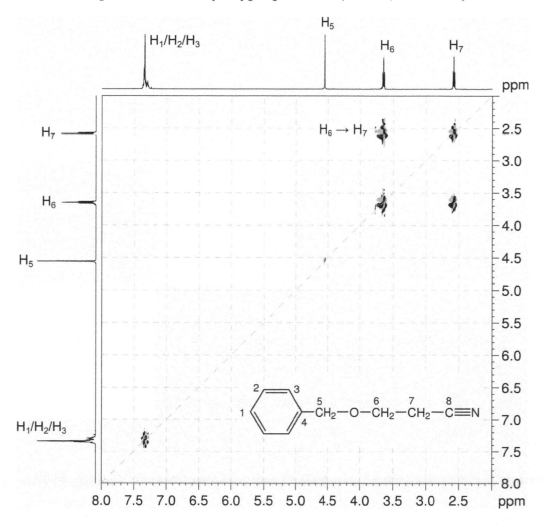

6. The protonated carbons C_5, C_6 and C_7 are easily assigned by the me-HSQC spectrum to resonances at 73.3, 64.4 and 18.9 ppm, respectively.

7. Although the aromatic H–C correlations in the me-HSQC spectrum are not resolved, the structure can still be determined.

^{1}H–^{13}C me-HSQC spectrum of benzyloxypropionitrile (CDCl$_3$, 500 MHz)

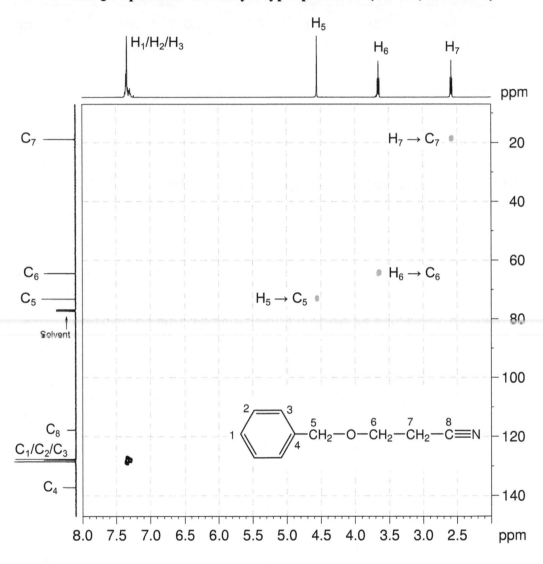

8. In the HMBC spectrum, the correlations from the aryl protons to the quaternary carbon signal at 137.3 ppm identifies the signal as the *ipso* carbon C_4.

9. The H_{Ar}–C_5, H_5–C_{Ar} and H_5–C_4 correlations in the HMBC spectrum place the C_5 CH_2 group next to the benzene ring.

10. There is no correlation between H_5 and H_6 in the COSY spectrum indicating that they are not adjacent to each other. However, the H_6–C_5 and H_5–C_6 correlations in the HMBC spectrum indicate that they are connected by a single atom, most likely oxygen as this would be consistent with the chemical shifts of H_5, H_6, C_5 and C_6.

11. We are left with one carbon atom (^{13}C resonance at 117.9 ppm, C_8), one nitrogen atom and two remaining degrees of unsaturation which, together with the IR data, are indicative of a nitrile group. The H_6–C_8 and H_7–C_8 correlations in the HMBC spectrum place the nitrile carbon next to methylene group C_7.

12. Note that in the HMBC spectrum there are unsuppressed one-bond correlations (H_5–C_5 and H_6–C_6) which appear as doublets.

^1H–^{13}C HMBC spectrum of benzyloxypropionitrile (CDCl$_3$, 500 MHz)

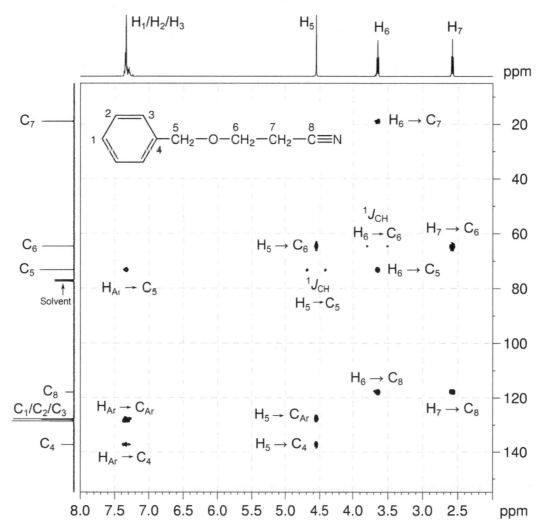

Problem 49

Question:

Identify the following compound.

Molecular Formula: $C_{10}H_{18}O$

Solution:

Cineole

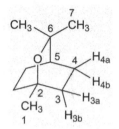

1. The molecular formula is $C_{10}H_{18}O$. Calculate the degree of unsaturation from the molecular formula: ignore the O atom to give an effective molecular formula of $C_{10}H_{18}$ (C_nH_m) which gives the degree of unsaturation as $(n - m/2 + 1) = 10 - 9 + 1 = 2$. The compound contains a combined total of two rings and/or π bonds.

2. Using a combination of the integration in the ^1H NMR spectrum and the me-HSQC spectrum, there are the following groups in the molecule:

    - three CH_3 groups at ^1H 1.05 / ^{13}C 27.6 ppm (H_1/C_1) and ^1H 1.24 / ^{13}C 28.9 ppm (H_7/C_7) – two of which are equivalent (H_7/C_7)

    - one CH group at ^1H 1.41 / ^{13}C 33.0 ppm (H_5/C_5)

    - two CH_2 groups with diastereotopic protons at ^1H 1.66, 1.50 / ^{13}C 31.6 ppm ($H_{3a},H_{3b}/C_3$) and ^1H 2.02, 1.50 / ^{13}C 22.9 ppm ($H_{4a},H_{4b}/C_4$) – based on integration, there are two sets of each CH_2 group.

3. From the ^{13}C NMR spectrum, there are also two quaternary carbons at 73.6 and 69.8 ppm (C_6 and C_2). These signals are shifted downfield and are likely bound to oxygen.

^1H NMR spectrum of cineole (CDCl₃, 600 MHz)

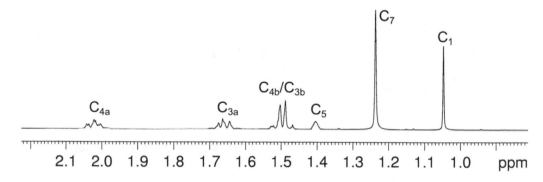

$^{13}C\{^1H\}$ NMR spectrum of cineole (CDCl₃, 150 MHz)

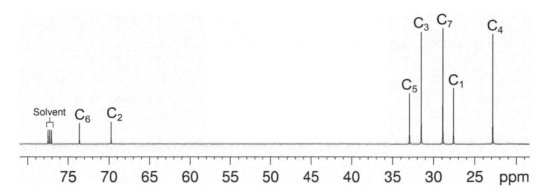

$^1H–^{13}C$ me-HSQC spectrum of cineole (CDCl₃, 600 MHz)

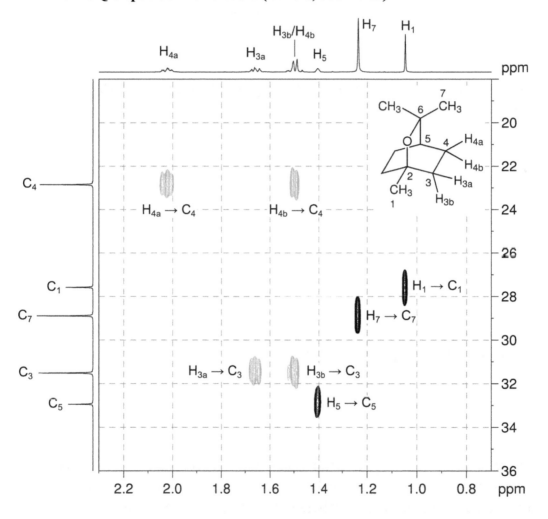

4. The INADEQUATE spectrum shows direct ^{13}C–^{13}C connectivity.

5. Beginning with the methyl carbon resonance at 27.6 ppm (C_1), the correlation to the quaternary carbon signal at 69.8 ppm (C_2) affords a CH_3–C fragment.

6. C_2 has a correlation to the methylene carbon signal at 31.6 ppm (C_3). There are two equivalent methylene groups for C_3 so we now have the following fragment:

$$CH_3 - C \Big< \begin{array}{l} CH_2 - \\ CH_2 - \end{array}$$

7. C_3 has a correlation to the methylene carbon signal at 22.9 ppm (C_4). There are two equivalent methylene groups for C_4 so we now have the following fragment:

$$CH_3 - C \Big< \begin{array}{l} CH_2 - CH_2 - \\ CH_2 - CH_2 - \end{array}$$

8. C_4 has a correlation to the methine carbon signal at 33.0 ppm (C_5) which extends the fragment to:

$$CH_3 - C \Big< \begin{array}{l} CH_2 - CH_2 \\ CH_2 - CH_2 \end{array} \Big> CH -$$

9. C_5 has a correlation to the quaternary carbon signal at 73.6 ppm (C_6) which affords:

$$CH_3 - C \Big< \begin{array}{l} CH_2 - CH_2 \\ CH_2 - CH_2 \end{array} \Big> CH - C \big<$$

10. C_6 has a correlation to the methyl carbon signal at 28.9 ppm (C_7). There are two equivalent methyl groups for C_7 which affords:

$$CH_3 - C \Big< \begin{array}{l} CH_2 - CH_2 \\ CH_2 - CH_2 \end{array} \Big> CH - C \Big< \begin{array}{l} CH_3 \\ CH_3 \end{array}$$

11. There remains a single oxygen atom in the molecular formula and it must be located between quaternary carbons C_2 and C_6 based on the chemical shifts of these carbons. There are no correlations in the INADEQUATE spectrum between C_2 and C_6 because of the oxygen atom between them.

$$CH_3 - C \Big< \begin{array}{l} CH_2 - CH_2 \\ CH_2 - CH_2 \end{array} \Big> O \quad CH - C \Big< \begin{array}{l} CH_3 \\ CH_3 \end{array}$$

12. Redrawing the structure affords:

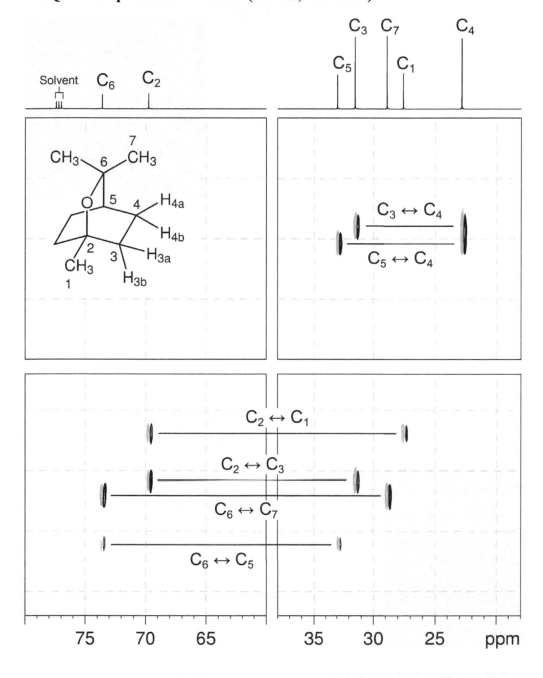

INADEQUATE spectrum of cineole (CDCl₃, 150 MHz)

Problem 50

Question:

Given below are nine benzoquinones which are isomers of $C_{10}H_{12}O_2$. The ^1H and $^{13}C\{^1H\}$ NMR spectra of one of the isomers are given below. The $^1H-^{13}C$ me-HSQC and $^1H-^{13}C$ HMBC spectra are given on the following page. To which of these compounds do the spectra belong?

HINT: In benzoquinones, three-bond C–H correlations in the HMBC spectrum ($^3J_{C-H}$) are significantly larger than two-bond C–H correlations ($^2J_{C-H}$) (which are negligible).

A   B   C   D

E   F   G   H   I

Solution:

Thymoquinone

(Isomer C)

1. The ^1H NMR spectrum shows that the alkene protons show only small couplings. The low-field signal at 6.60 ppm is a quartet (long-range coupling to a –CH$_3$ group), and the high-field signal at 6.53 ppm is a doublet (long-range coupling to a –CH group). There is no evidence that the two alkene protons are mutually coupled, so isomers **A**, **F**, **H** and **I** may be eliminated as possibilities.

^{1}H NMR spectrum of thymoquinone (CDCl$_3$, 500 MHz)

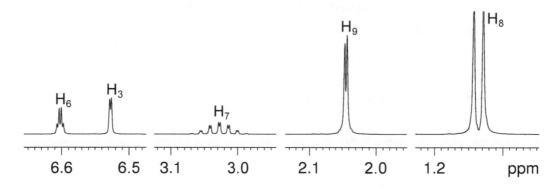

^{13}C{^{1}H} NMR spectrum of thymoquinone (CDCl$_3$, 150 MHz)

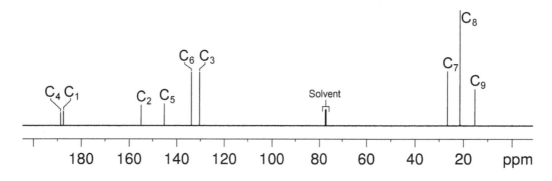

2. The ^{1}H–^{13}C me-HSQC spectrum shows direct (one-bond) correlations between proton and carbon nuclei. The protonated carbons are easily identified as signals at 133.9, 130.4, 26.6, 21.4 and 15.4 ppm.

^{1}H–^{13}C me-HSQC spectrum of thymoquinone (CDCl$_3$, 600 MHz)

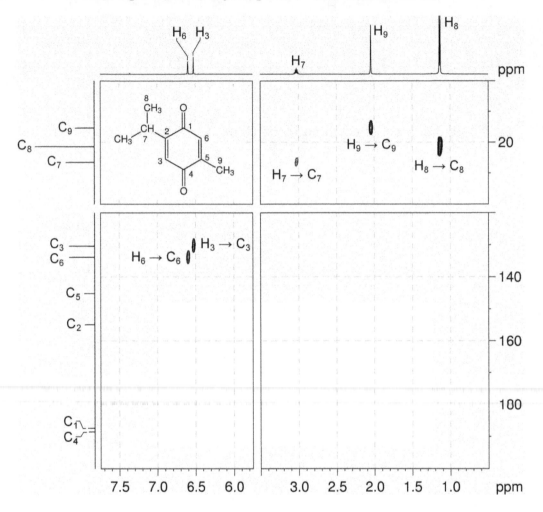

3. In the HMBC spectra of benzoquinones, we expect to see the strongest cross-peaks for three-bond correlations, rather than two-bond correlations (which are negligible in size, and therefore not observed in HMBC spectra).

^1H–^{13}C HMBC spectrum of thymoquinone (CDCl$_3$, 600 MHz)

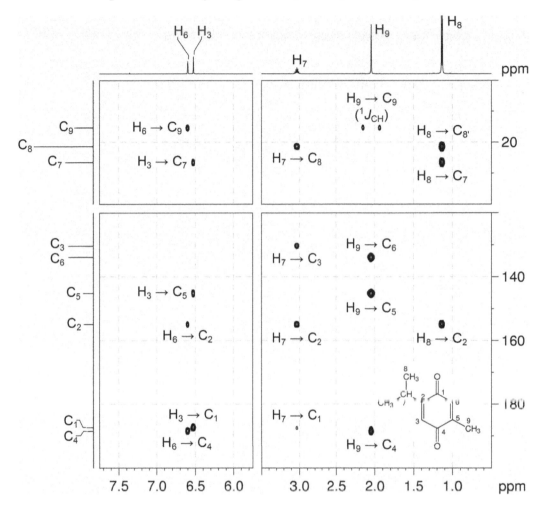

4. Each alkene proton correlates to a different carbonyl carbon resonance. In the HMBC spectrum of isomers B, D and E the alkene protons would correlate to the same carbonyl carbon; therefore, these isomers may be eliminated. Only isomers C and G remain as possibilities.

5. The quartet coupling pattern in the ^1H NMR spectrum identifies the low-field alkene resonance (at 6.60 ppm) as being adjacent to the methyl group, while the high-field alkene resonance (a doublet at 6.52 ppm) is adjacent to the isopropyl group. These assignments are confirmed in the HMBC spectrum (H_6–C_9 and H_3–C_7 correlations).

6. There are correlations from H_9 (the –CH$_3$ protons) and the adjacent alkene proton H_6 to the low-field carbonyl resonance (C_4). There are also correlations from H_7 (the isopropyl

–CH proton) and the adjacent alkene proton H_3 to the high-field carbonyl resonance (C_1). In isomer C, the H_9–C_4 and H_7–C_1 correlations would both arise from three-bond couplings, while in isomer G, these correlations must arise from five-bond coupling (as the ring-substituent and its "adjacent" alkene proton are correlated to the same carbonyl carbon resonance). Five-bond couplings are very rare, so isomer G may be eliminated, and **isomer C** is the correct answer.

7.   There is also a correlation between H_8 and $C_{8'}$. While this appears to be a one-bond correlation, in *gem*-dimethyl groups, the apparent one-bond correlation arises from the $^3J_{C-H}$ interaction of the protons of one of the methyl groups with the chemically equivalent carbon which is three bonds away.

Problem 51

Question:

Identify the following compound.

Molecular Formula: C_9H_9BrO

IR: 3164 (br), 2953 cm^{-1}.

Solution:

4-Bromo-1-indanol

1. The molecular formula is C_9H_9BrO. Calculate the degree of unsaturation from the molecular formula: replace the Br by H and ignore the O atom to give an effective molecular formula of C_9H_{10} (C_nH_m) which gives the degree of unsaturation as $(n - m/2 + 1) = 9 - 5 + 1 = 5$. The compound contains a combined total of five rings and/or π bonds.

2. 1D spectra establish that the compound is a trisubstituted benzene which accounts for four degrees of unsaturation. There remains one additional degree of unsaturation so there must be one additional ring or double bond.

3. The coupling pattern in the aromatic region of the ^{1}H NMR spectrum shows that the three protons on the aromatic ring must occupy adjacent positions. The proton at 7.10 ppm is the proton at the middle of the spin system since it has two large splittings (*i.e.* it has two *ortho*-protons) whereas the other two aromatic protons have only one large splitting.

4. In the alkyl region of the ^{1}H NMR spectrum, there is one exchangeable proton at 2.12 ppm (which must be bound to the oxygen to form an alcohol) and there are five signals each of integration one.

5. There are three signals in the alkyl region of the ^{13}C{^{1}H} NMR spectrum (including one that is obscured by the solvent signal), so some of the alkyl protons must be diastereotopic. This indicates that a chiral centre is likely to be present in the molecule.

¹H NMR spectrum of 4-bromo-1-indanol (CDCl₃, 500 MHz)

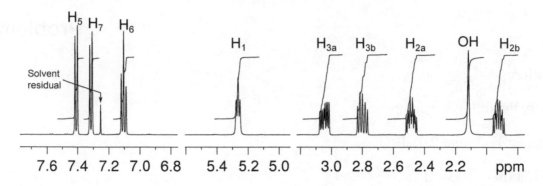

¹³C{¹H} NMR spectrum of 4-bromo-1-indanol (CDCl₃, 125 MHz)

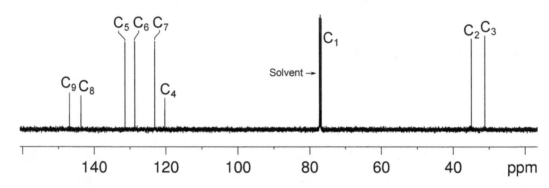

6. The ^1H–^{13}C me-HSQC spectrum can be used to identify pairs of diastereotopic protons, as they will correlate to the same ^{13}C signal. The signals at 3.05 and 2.81 ppm constitute one diastereotopic methylene group (H_{3a} and H_{3b}), with the associated carbon (C_3) at 31.2 ppm; a second methylene group resonates at 2.48 and 1.93 ppm (H_{2a} and H_{2b}), with the associated carbon (C_2) at 34.9 ppm. The signal at 5.26 ppm corresponds to a methine (CH) group (H_1), and C_1 resonates at 77.1 ppm.

^1H–^{13}C me-HSQC spectrum of 4-bromo-1-indanol (CDCl$_3$, 500 MHz)

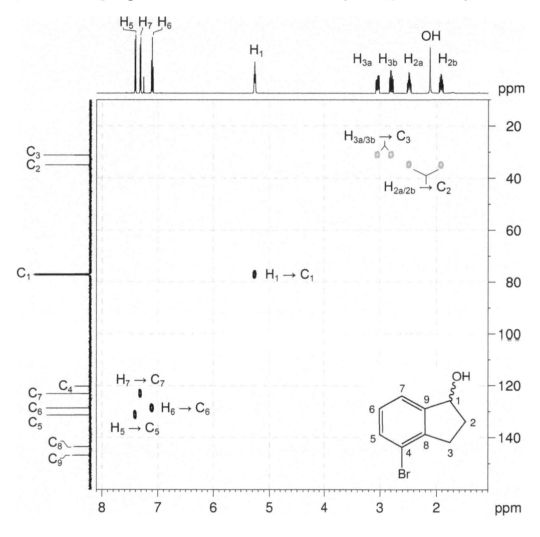

7. Listing the currently known fragments gives:

8. The ^{1}H–^{1}H COSY spectrum identifies the coupling relationship between the alkyl groups – H_1 couples to H_{2a} and H_{2b}, and H_{2a} and H_{2b} both couple to H_{3a} and H_{3b}. H_1 also couples to the alcohol. We therefore have:

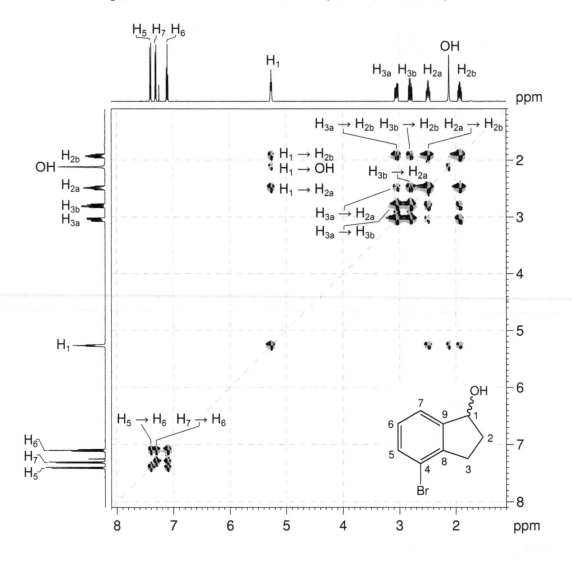

^{1}H–^{1}H COSY spectrum of 4-bromo-1-indanol (CDCl$_3$, 500 MHz)

9. There are two possible compounds that can be constructed from the available fragments:

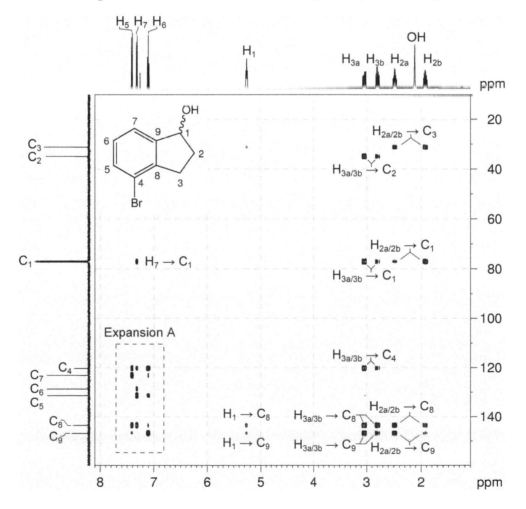

A   B

10. In the $^1H–^{13}C$ HMBC spectrum, there is a correlation between one of the "outer" aryl protons (at 7.31 ppm) and the methine carbon at 77.1 ppm (C_1). Remember that in aromatic systems, it is the three-bond coupling that is the largest, both around the ring and to benzylic carbon atoms, so this correlation suggests that there is an aromatic proton three bonds removed from C_1. This eliminates Isomer A as a possibility, and identifies **Compound B** as the correct isomer (4-bromo-1-indanol). It also identifies the aromatic resonance at 7.31 ppm as the proton closest to the methine group.

11. We can then assign all the aromatic resonances in the ^1H and ^{13}C spectra using the $^1H–^{13}C$ me-HSQC and HMBC spectra and confirm all spectra are consistent with the structure.

$^1H–^{13}C$ HMBC spectrum of 4-bromo-1-indanol (CDCl$_3$, 500 MHz)

^{1}H–^{13}C HMBC spectrum of 4-bromo-1-indanol – expansion A

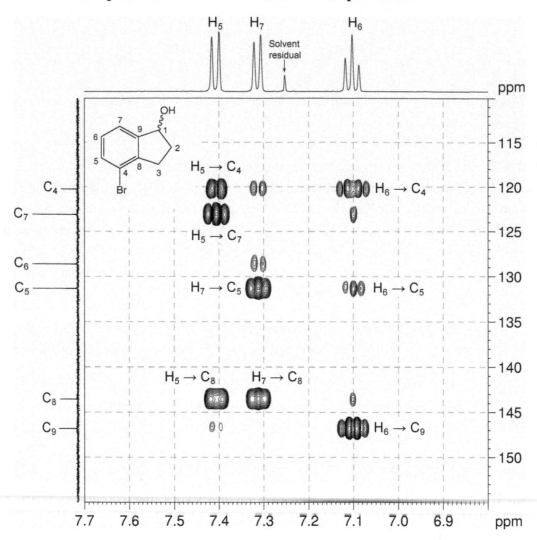

Question:

The ^1H and $^{13}C\{^1H\}$ NMR spectra of 1-bromo-4-methylnaphthalene ($C_{11}H_9Br$) recorded in $CDCl_3$ solution at 298 K and 500 MHz are given below.

The ^1H NMR spectrum has signals at δ 2.58, 7.08, 7.50, 7.54, 7.61, 7.91 and 8.23 ppm.

The $^{13}C\{^1H\}$ NMR spectrum has signals at δ 19.2, 120.6, 124.5, 126.4, 126.85, 126.91, 127.6, 129.4, 131.7, 133.7 and 134.3 ppm.

The 2D ^1H–^1H COSY, multiplicity-edited ^1H–^{13}C HSQC and ^1H–^{13}C HMBC spectra are given on the following pages. Use these spectra to assign the ^1H and $^{13}C\{^1H\}$ resonances for this compound.

Solution:

1-Bromo-4-methylnaphthalene

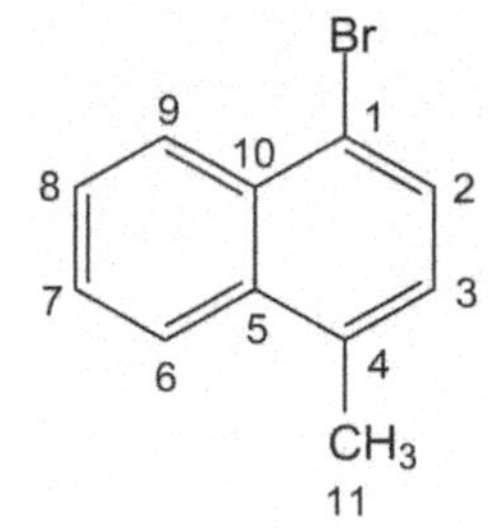

Proton	Chemical Shift (ppm)	Carbon	Chemical Shift (ppm)
		C_1	120.6
H_2	7.61	C_2	129.5
H_3	7.08	C_3	126.91
		C_4	134.3
		C_5	133.7
H_6	7.91	C_6	124.5
H_7	7.50	C_7	126.4
H_8	7.54	C_8	126.85
H_9	8.23	C_9	127.6
		C_{10}	131.7
H_{11}	2.58	C_{11}	19.2

1. H_{11} and C_{11} are easily identified on the basis of chemical shift.

^{1}H NMR spectrum of 1-bromo-4-methylnaphthalene (CDCl$_3$, 500 MHz)

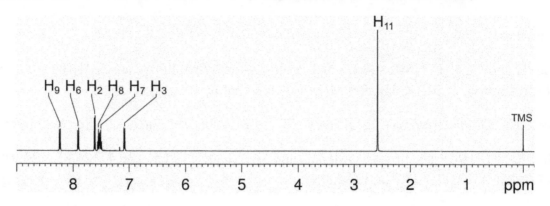

^{13}C{^{1}H} NMR spectrum of 1-bromo-4-methylnaphthalene – expansion (CDCl$_3$, 125 MHz)

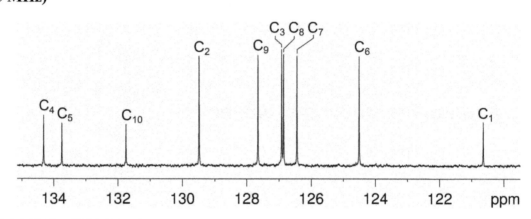

2. In the COSY spectrum, there is a correlation between H_{11} and the aromatic proton at 7.08 ppm. This is a four-bond correlation between H_{11} and H_3. Careful analysis of the me-HSQC spectrum (expansion B) identifies C_3 as the resonance at 126.91 ppm.

^{1}H–^{1}H COSY spectrum of 1-bromo-4-methylnaphthalene (CDCl$_3$, 500 MHz)

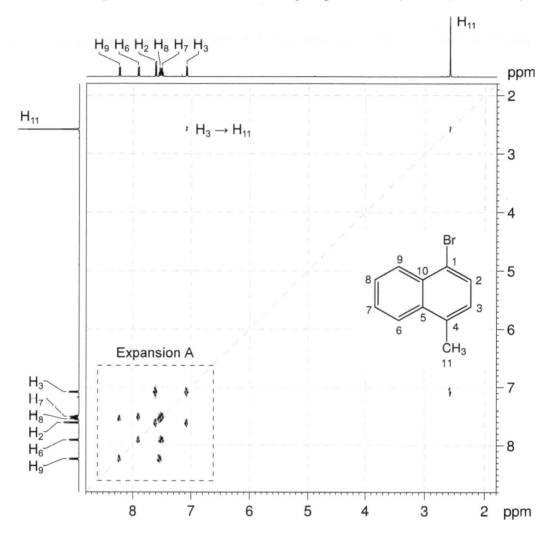

^1H–^{13}C me-HSQC of 1-bromo-4-methylnaphthalene – expansion B (CDCl$_3$, 500 MHz)

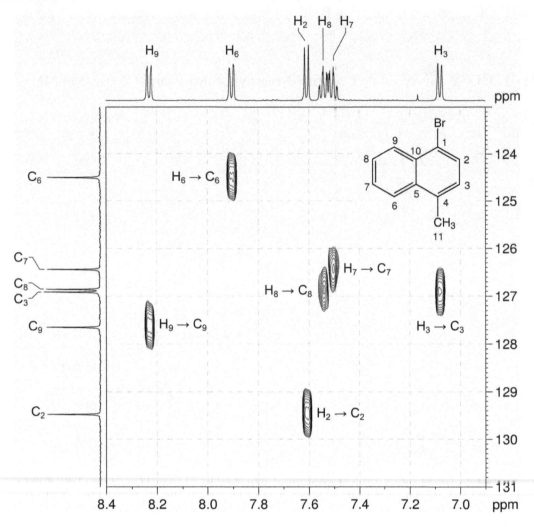

3. In the COSY spectrum (expansion A), H_3 correlates to the resonance at 7.61 ppm, and this resonance must be due to H_2. The me-HSQC spectrum then identifies C_2 as the resonance at 129.5 ppm.

^1H–^1H COSY spectrum of 1-bromo-4-methylnaphthalene – expansion A

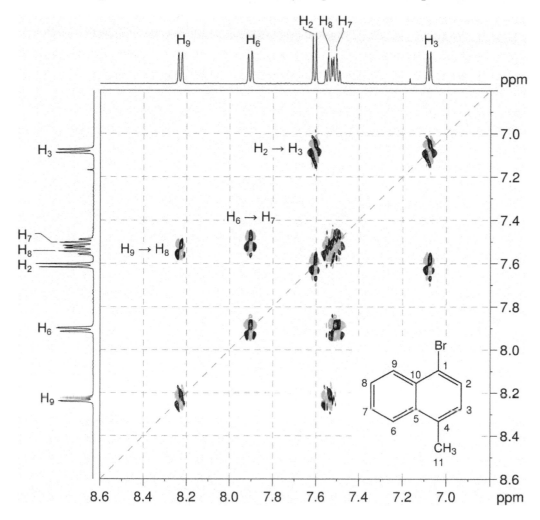

4. Based on the coupling pattern, the ^1H signals at 8.23 and 7.91 ppm are due to either H_6 or H_9 while the signals at 7.54 and 7.50 ppm are due to either H_7 or H_8.

5. Remember that, in aromatic systems, the three-bond coupling $^3J_{C-H}$ is typically the larger long-range coupling and gives rise to the strongest cross-peaks in HMBC spectra.

6. In expansion C of the HMBC spectrum, the low-field ^{13}C signal at 134.3 ppm has correlations from the ^1H signals at 7.61 ppm (previously assigned to H_2) and 7.91 ppm. The ^{13}C signal at 134.3 ppm is due to either C_4 or C_{10} which are both three bonds away from H_2. If the signal was due to C_{10} we would expect a correlation from H_8, and as such a correlation is absent, the signal must be due to C_4.

7. Therefore the ^1H resonance at 7.91 ppm must be due to H_6 which is three bonds away from C_4. The me-HSQC spectrum can now be used to assign C_6 to the ^{13}C signal at 124.5 ppm.

^1H–^{13}C HMBC of 1-bromo-4-methylnaphthalene – expansion C (CDCl₃, 500 MHz)

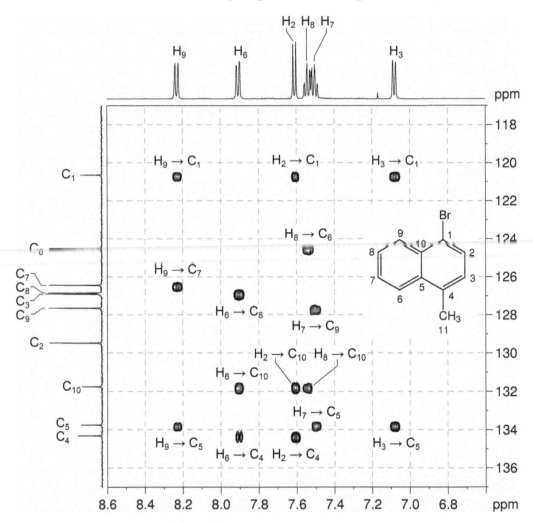

8. As H_6 has been assigned to the ^1H signal at 7.91 ppm, the signal at 8.23 ppm must be due to H_9. C_9 can be assigned to the ^{13}C signal at 127.6 ppm using the me-HSQC spectrum.

9. From the COSY spectrum, H_6 correlates to the signal at 7.50 ppm which can be assigned to H_7. C_7 can be assigned to the ^{13}C signal at 126.4 ppm using the me-HSQC spectrum.

10. In expansion C of the HMBC spectrum, the ^{13}C signal at 133.7 ppm has correlations from ^1H signals at 8.23 (H_9), 7.50 (H_7) and 7.08 (H_3) ppm and must be due to C_5.

11. As H_7 has been assigned to the ^1H signal at 7.50 ppm, H_8 can be assigned to the signal at 7.54 ppm. C_8 can be assigned to the ^{13}C signal at 126.85 ppm using the me-HSQC spectrum.

12. In the HMBC spectrum (expansion C), the correlations from H_6, H_2 and H_8 to the ^{13}C signal at 131.7 ppm identifies the resonance as due to C_{10}.

13. In the HMBC spectrum (expansion C), the H_9, H_2 and H_3 correlations to the ^{13}C signal at 120.6 ppm identifies the resonance as due to C_1.

14. Note that there is an unusually long-range (five-bond) coupling from H_{11} to C_1 in expansion D of the HMBC spectrum.

^1H–^{13}C HMBC of 1-bromo-4-methylnaphthalene – expansion D

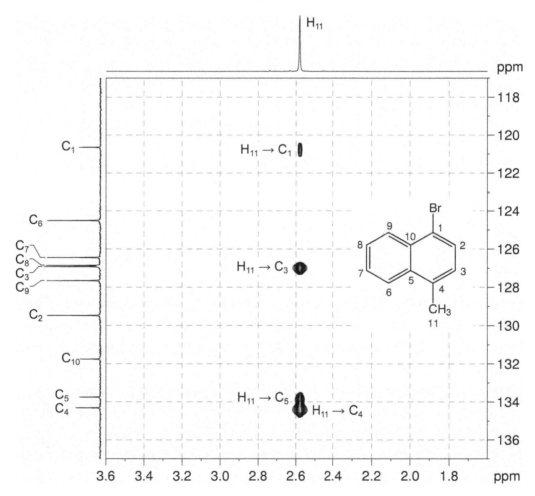

Problem 53

Question:

Identify the following compound.

Molecular Formula: $C_{10}H_{14}O$

IR: 3377 (br) cm^{-1}

Solution:

Carvacrol

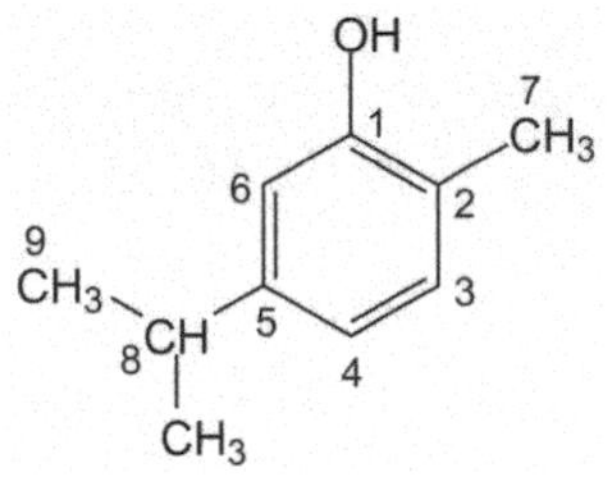

1. The molecular formula is $C_{10}H_{14}O$. Calculate the degree of unsaturation from the molecular formula: ignore the O atom to give an effective molecular formula of $C_{10}H_{14}$ (C_nH_m) which gives the degree of unsaturation as $(n - m/2 + 1) = 10 - 7 + 1 = 4$. The compound contains a combined total of four rings and/or π bonds.

2. 1D NMR data establish that the compound is a trisubstituted benzene with an –OH substituent (^{1}H resonance at 4.86 ppm, exchangeable), a –CH$_3$ substituent (^{1}H resonance at 2.21 ppm) and an isopropyl substituent (septet at 2.80 ppm and doublet at 1.20 ppm in ^{1}H NMR spectrum). The benzene ring accounts for all of the degrees of unsaturation, so the compound contains no additional rings or multiple bonds.

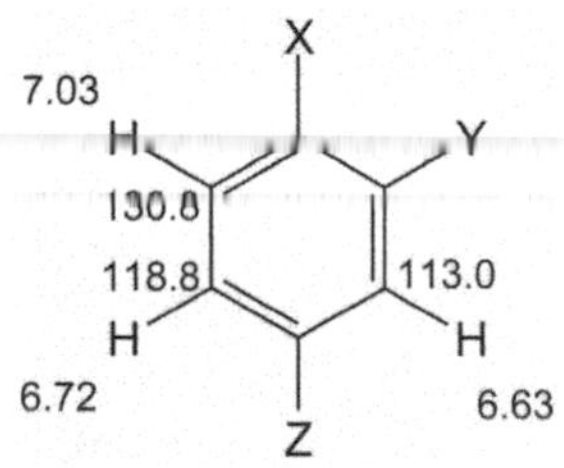

3. The coupling pattern in the aromatic region of the ^{1}H NMR spectrum shows that the compound must be a "1,2,4-trisubstituted benzene". The proton at 6.63 ppm clearly has no *ortho* couplings and must be the isolated proton in the spin system sandwiched between two substituents. The two protons at 6.72 and 7.03 ppm each have a large *ortho* coupling so they must be adjacent to each other. The proton at 6.72 ppm has an additional *meta* coupling so it must be the proton *meta* to the proton at 6.63 ppm. The remaining proton at 7.03 ppm must be the proton *para* to the proton at 6.63 ppm.

^{1}H NMR spectrum of carvacrol (CDCl$_3$, 400 MHz)

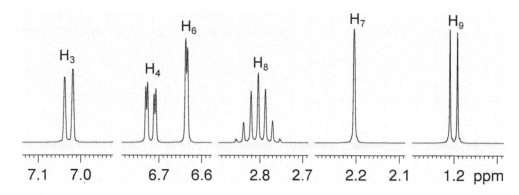

^{13}C{^{1}H} NMR spectrum of carvacrol (CDCl$_3$, 400 MHz)

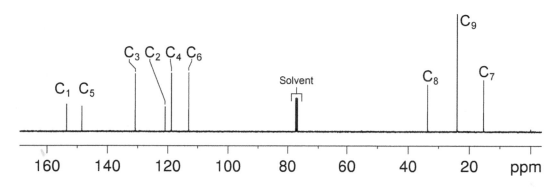

4.   In the COSY spectrum (expansion A), there are correlations from the aromatic protons at 7.03 and 6.63 ppm to the methyl protons at 2.21 ppm. However we are not able to deduce the position of the methyl group on the aromatic ring based on these correlations.

^{1}H–^{1}H COSY spectrum of carvacrol – expansion A (CDCl$_3$, 400 MHz)

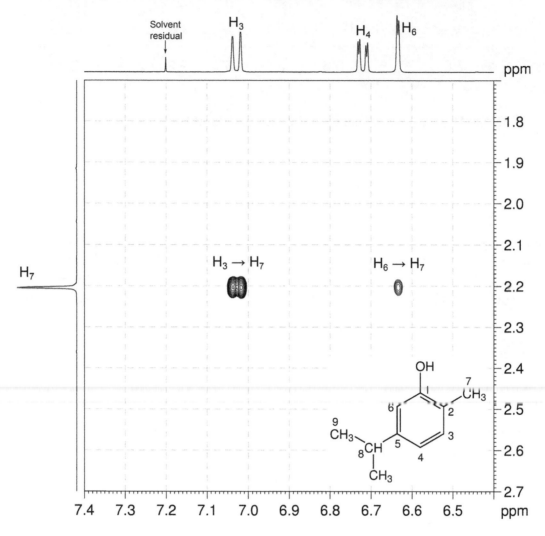

5. Expansion B of the COSY spectrum shows the expected correlations between the aromatic protons.

^{1}H–^{1}H COSY spectrum of carvacrol – expansion B

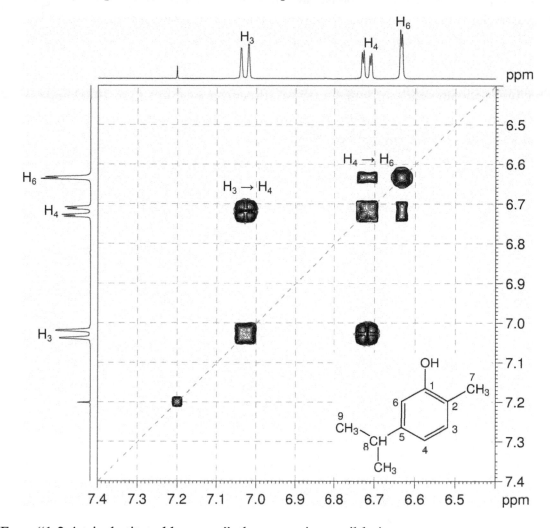

6. For a "1,2,4-trisubstituted benzene", there are six possible isomers:

A   B   C

D   E   F

7. The ^1H–^{13}C me-HSQC spectrum easily identifies the protonated carbons. The isolated methyl carbon is at 15.3 ppm while the methine and methyl carbons of the isopropyl group are at 33.7 and 24.0 ppm, respectively. The isolated aromatic proton at 6.63 ppm correlates to the carbon at 113.0 ppm. The aromatic protons at 6.72 and 7.03 ppm correlate to the carbons at 118.8 and 130.8 ppm, respectively

8. The three aromatic carbons which bear substituents appear at 153.5, 148.4 and 120.9 ppm. The low-field carbon resonance (153.3 ppm) must be the carbon bearing the –OH substituent based on its chemical shift.

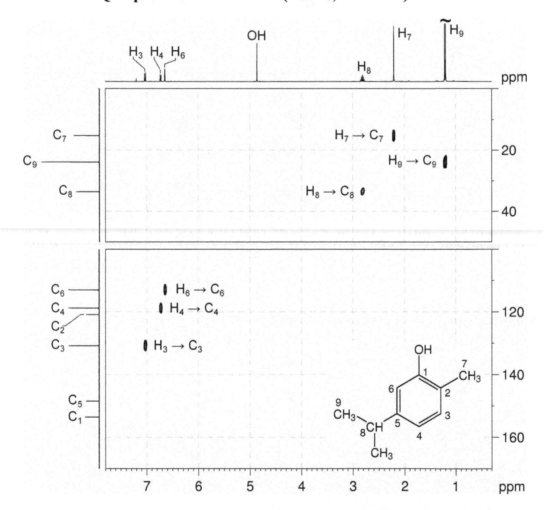

^1H–^{13}C me-HSQC spectrum of carvacrol (CDCl$_3$, 400 MHz)

9. Remember that, in aromatic systems, the three-bond coupling $^3J_{C–H}$ is typically the larger long-range coupling and gives rise to the strongest cross-peaks. There are also strong correlations between benzylic carbon atoms and protons which are located *ortho* to the benzylic substituent.

10. In the ^{1}H–^{13}C HMBC spectrum, the aromatic proton at 7.03 ppm correlates to the methyl carbon at 15.3 ppm, placing the isolated methyl group *ortho* to the proton at 7.03 ppm (at position X on the aromatic ring). This eliminates Isomers A, B, C and E.

11. Both aromatic protons at 6.72 and 6.63 ppm correlate to the isopropyl CH carbon at 33.7 ppm, which places the isopropyl group at position Z on the aromatic ring. This eliminates Isomer D, and identifies **Isomer F** as the correct isomer.

^{1}H–^{13}C HMBC spectrum of carvacrol (CDCl₃, 400 MHz)

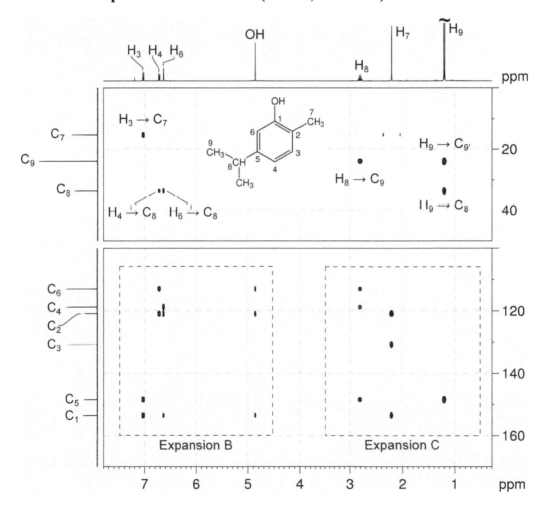

^{1}H–^{13}C HMBC spectrum of carvacrol – expansion B

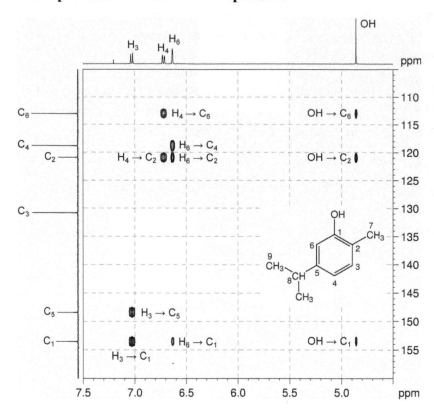

^{1}H–^{13}C HMBC spectrum of carvacrol – expansion C

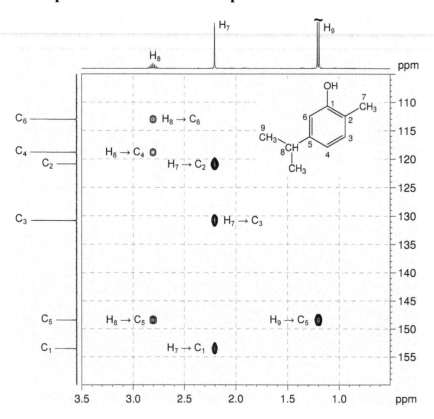

12. Verify that all spectroscopic data are consistent with the structure.

13. In the HMBC spectrum, the methyl protons (H_7) correlate to C_1 and C_3 (three-bond correlations) as well as to the *ipso* carbon (C_2, a two-bond correlation, identified as the signal at 120.9 ppm).

14. Similarly, the isopropyl methine proton (H_8) correlates to C_4 and C_6 (three-bond correlations) as well as the *ipso* carbon (C_5, a two-bond correlation, identified as the signal at 148.4 ppm).

15. In expansion C of the HMBC spectrum, the isopropyl methyl protons (H_9) correlate only to the aromatic *ipso* carbon (C_5, a three-bond correlation).

16. In the HMBC spectrum, the –OH proton also correlates to the *ipso* carbon C_1 (a two-bond correlation), and C_2 and C_6, which are three bonds away. Note that correlations to exchangeable protons are not always observed.

17. Note also that in the HMBC spectrum for this compound, there is a strong correlation between H_9 and $C_{9'}$. While this appears to be a one-bond correlation, in isopropyl groups, the apparent one-bond correlation arises from the $^3J_{C-H}$ interaction of the protons of one of the methyl groups with the chemically equivalent carbon which is three bonds away.

Question:

This compound is an amide, and amides frequently exist as a mixture of stereoisomers due to restricted rotation about the C–N bond.

Use the spectra below to identify the compound, including the stereochemistry of the major isomer.

Molecular Formula: $C_{10}H_{11}NO_2$

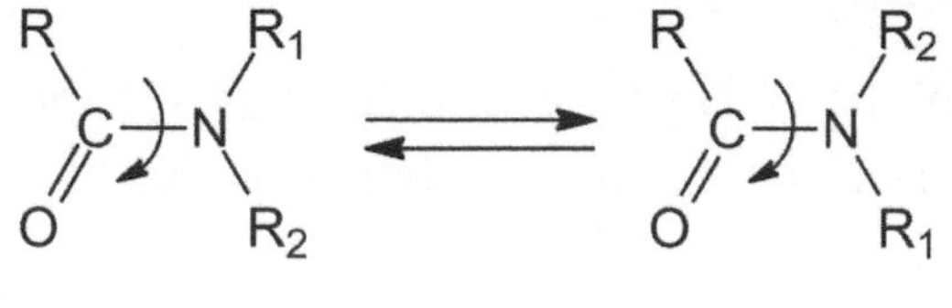

Solution:

Acetoacetanilide

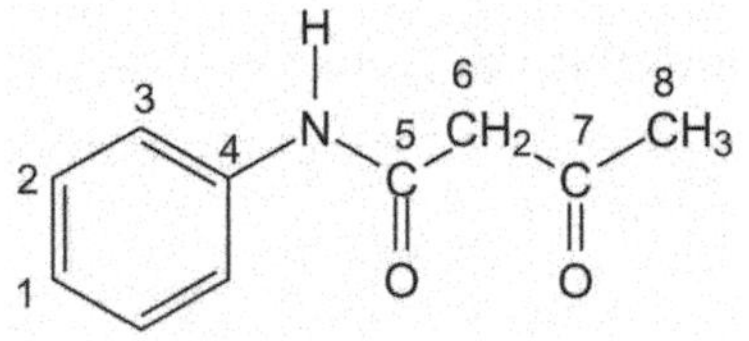

1. The molecular formula is $C_{10}H_{11}NO_2$. Calculate the degree of unsaturation from the molecular formula: ignore the O atoms and ignore the N and remove one H to give an effective molecular formula of $C_{10}H_{10}$ (C_nH_m) which gives the degree of unsaturation as $(n - m/2 + 1) = 10 - 5 + 1 = 6$. The compound contains a combined total of six rings and/or π bonds.

2. 1D NMR data establish that the compound is a mono-substituted benzene ring (^1H resonances at 7.60, 7.32 and 7.06 ppm for H_3, H_2 and H_1, respectively), contains an amide group (^1H resonance at 10.09 ppm which is exchangeable on warming, also ^{13}C resonance at 165.0 ppm – C_5) and a ketone functional group (^{13}C resonance at 202.8 ppm, C_7). This accounts for all of the degrees of unsaturation, so the compound contains no additional rings or multiple bonds.

3. 1D NMR data also establish the presence of a CH_2 group (^1H resonance at 3.57 ppm, H_6) and a CH_3 group (^1H resonance at 2.22 ppm, H_8).

¹H NMR spectrum of acetoacetanilide (DMSO-*d*₆, 300 MHz)

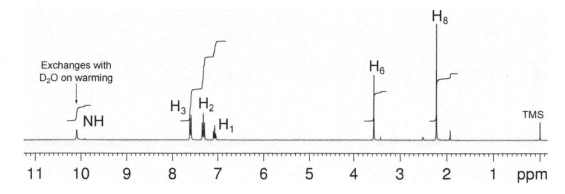

¹³C{¹H} NMR spectrum of acetoacetanilide (DMSO-*d*₆, 75 MHz)

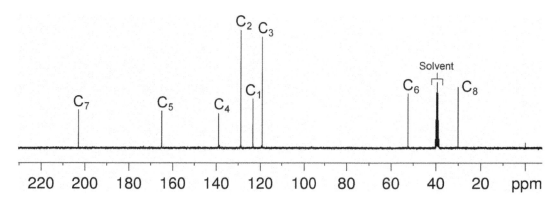

4. The protonated carbons C_2, C_1, C_3, C_6 and C_8 are easily identified by the me-HSQC spectrum as ^{13}C signals at 128.7, 123.4, 119.1, 52.3 and 30.2 ppm, respectively.

^1H–^{13}C me-HSQC spectrum of acetoacetanilide (DMSO-d_6, 300 MHz)

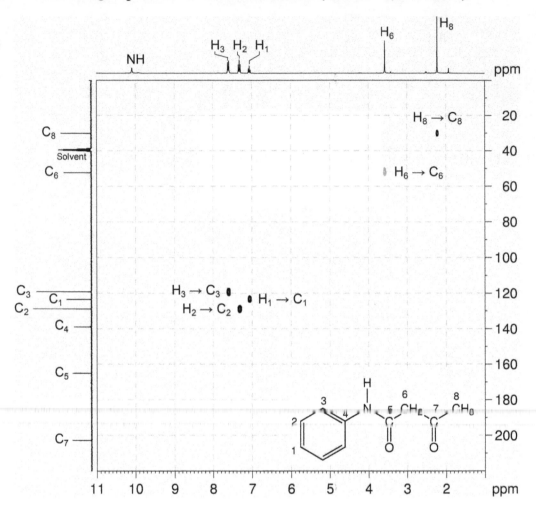

5. In the HMBC spectrum, the H_6–C_7 and H_8–C_7 correlations indicate that the CH_2 and CH_3 groups are bound to the ketone.

6. In the HMBC spectrum, the H_6–C_5 correlation indicates the CH_2 group is also bound to the amide group. As there is no coupling between the NH and CH_2 groups, the CH_2 group must be bound to the amide carbon.

7. The NH–C_3 correlation in the HMBC spectrum indicates the NH is bound directly to the aromatic ring.

^{1}H–^{13}C HMBC spectrum of acetoacetanilide (DMSO-d_6, 300 MHz)

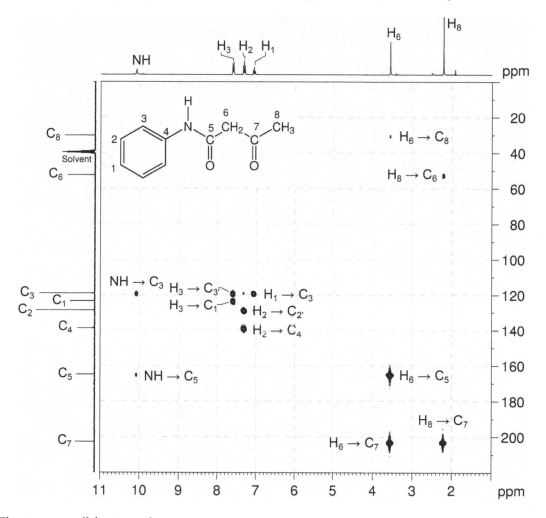

8. The two possible stereoisomers are:

A   B

9. In the NOESY spectrum, the strong NH–H$_6$ correlation identifies the correct stereoisomer as isomer **A**.

^{1}H–^{1}H NOESY spectrum of acetoacetanilide (DMSO-d_6, 300 MHz)

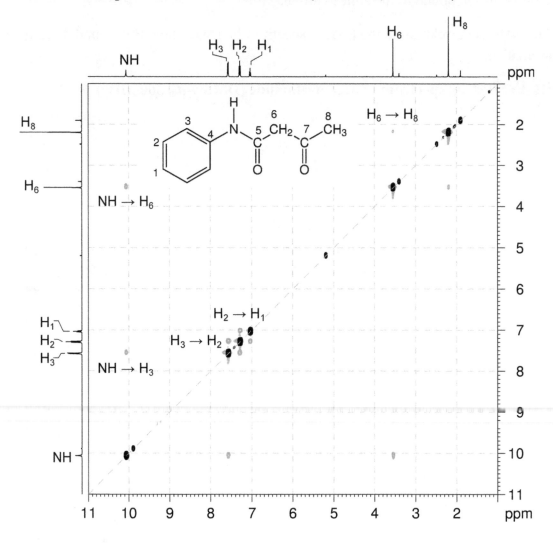

Question:

Identify the following compound.

Molecular Formula: $C_7H_{10}N_2O_3$

IR: 2930, 2258, 1740, 1698, 1643 cm^{-1}

Solution:

Ethyl acetamidocyanoacetate

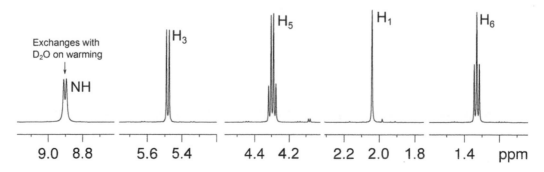

1. The molecular formula is $C_7H_{10}N_2O_3$. Calculate the degree of unsaturation from the molecular formula: ignore the O atoms, ignore the N atoms and remove two H atoms to give an effective molecular formula of C_7H_8 (C_nH_m) which gives the degree of unsaturation as $(n - m/2 + 1) = 7 - 4 + 1 = 4$. The compound contains a combined total of four rings and/or π bonds.

2. 1D NMR data establish the presence of an amide group (^1H resonance at 8.90 ppm which is exchangeable on warming) and two carbonyl groups (^{13}C resonances at 170.3 and 163.8 ppm for C_2 and C_4, respectively) – one belonging to the amide group and the other possibly an ester group based on chemical shift. The two carbonyl functional groups account for two degrees of unsaturation, so the compound contains an additional two rings or multiple bonds. The remaining quaternary carbon signal at 114.8 ppm (C_7) may be due to a CN group based on chemical shift and IR signal at 2258 cm^{-1}.

3. 1D NMR and me-HSQC spectra also establish the presence of a CH group (^1H 5.48 ppm / ^{13}C 43.4 ppm, H_3/C_3), a CH_2 group (^1H 4.30 ppm / ^{13}C 63.3 ppm, H_5/C_5) and two CH_3 groups (^1H 2.04 ppm / ^{13}C 22.1 ppm, H_1/C_1; ^1H 1.33 ppm / ^{13}C 13.9 ppm, H_6/C_6). The CH_2 group must be bound to oxygen on the basis of its chemical shift.

^1H NMR spectrum of ethyl acetamidocyanoacetate (CDCl$_3$/DMSO-d_6, 500 MHz)

$^{13}C\{^1H\}$ NMR spectrum of ethyl acetamidocyanoacetate (CDCl₃/DMSO-d_6, 125 MHz)

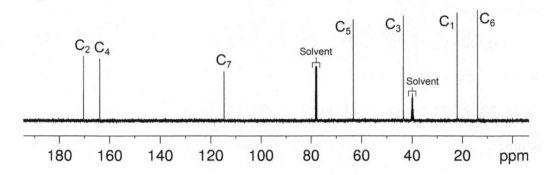

^1H–^{13}C me-HSQC spectrum of ethyl acetamidocyanoacetate (CDCl₃/DMSO-d_6, 500 MHz)

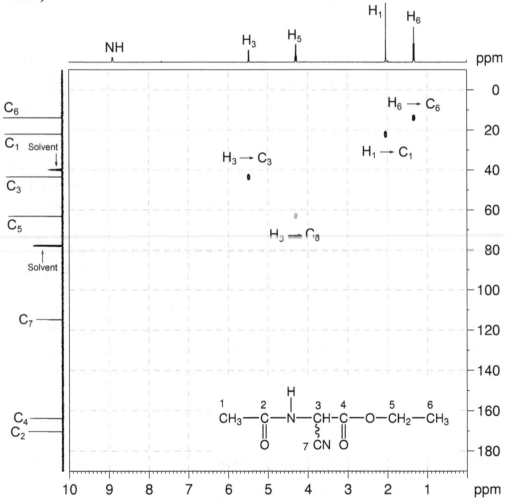

4. In the COSY spectrum, there are two separate spin systems – one for an ethyl group (H$_5$–H$_6$ correlation) and the other for the NH–CH group (NH–H$_3$ correlation).

^{1}H–^{1}H COSY spectrum of ethyl acetamidocyanoacetate (CDCl$_3$/DMSO-d_6, 500 MHz)

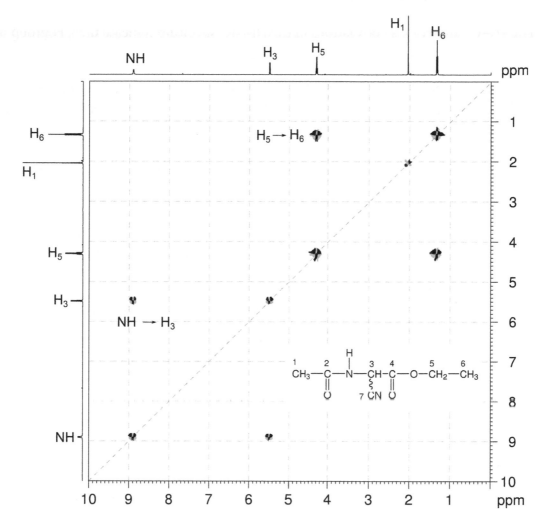

5.   In the HMBC spectrum, the NH correlation to the ^{13}C signal at 170.3 ppm identifies this resonance as due to the amide carbon C_2.

6.   The H_1–C_2 correlation in the HMBC spectrum indicates the CH_3 group is bound to the amide carbonyl group. It cannot be bound to the amide nitrogen as there is no coupling between the H_1 and NH as there is between NH and H_3 (of the CH group). This affords the $CH_3C(O)NHCH$– fragment.

7.   The H_3–C_7 and H_3–C_4 correlations in the HMBC spectrum indicate the CH group is bonded to both quaternary carbon C_7 and carbonyl carbon C_4.

8.   Carbonyl carbon C_4 must be part of an ester group as the H_5–C_4 correlation in the HMBC spectrum as well as the chemical shift of H_5/C_5 indicates the presence of an ethyl ester. All oxygen atoms in the molecule are now accounted for.

9.   The remaining N atom, two degrees of unsaturation plus IR data and chemical shift of C_7 points to a nitrile group at C_7.

^1H–^{13}C HMBC spectrum of ethyl acetamidocyanoacetate (CDCl₃/DMSO-d_6, 500 MHz)

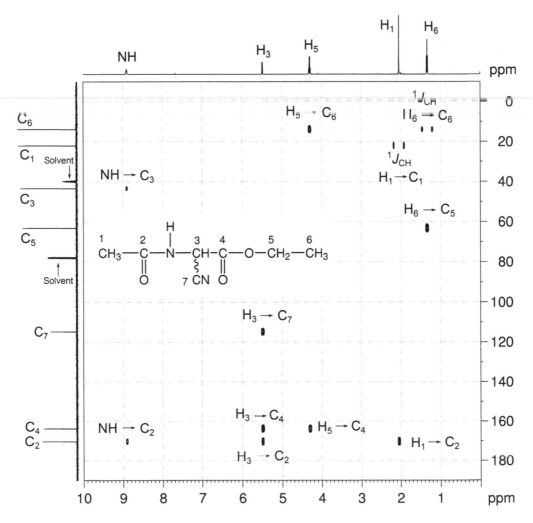

10. The ^1H–^{15}N HSQC spectrum shows one-bond correlations between proton and nitrogen nuclei. In the ^1H–^{15}N HSQC spectrum, the NH proton signal at 9.0 ppm correlates to the NH nitrogen signal at 111 ppm.

11. Note that the nitrile nitrogen does not appear in the ^1H–^{15}N HSQC spectrum since there are no protons attached to N.

^1H–^{15}N HSQC spectrum of ethyl acetamidocyanoacetate (CDCl₃/DMSO-d_6, 600 MHz)

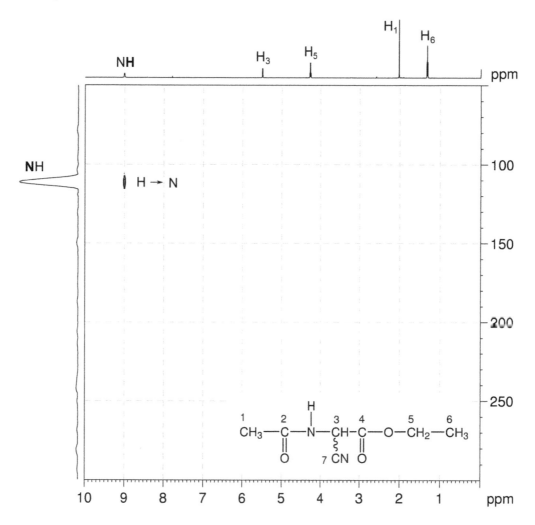

12. The ^{1}H–^{15}N HMBC spectrum shows two- and three-bond correlations between proton and nitrogen nuclei. In the ^{1}H–^{15}N HMBC spectrum, the H$_1$ proton signal at 2.0 ppm correlates to the NH nitrogen signal at 111 ppm whilst the H$_3$ proton signal at 5.5 ppm correlates to both the NH nitrogen signal at 111 ppm and the CN nitrogen signal at 251 ppm.

^{1}H–^{15}N HMBC spectrum of ethyl acetamidocyanoacetate (CDCl$_3$/DMSO-d_6, 600 MHz)

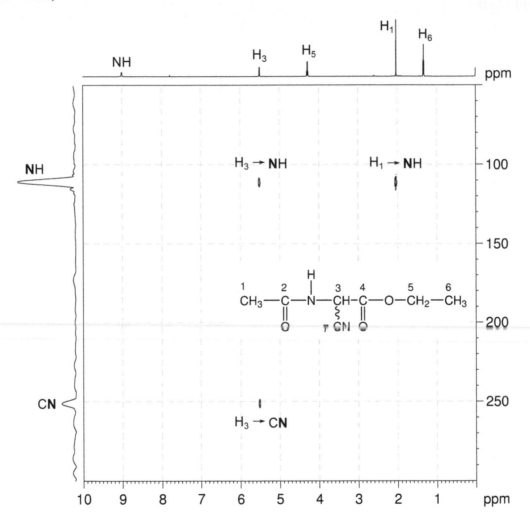

Question:

The following spectra belong to a macrocyclic sesquiterpene which contains three *trans*-substituted double bonds. Identify the compound.

Molecular Formula: $C_{15}H_{24}$

Solution:

α-Humulene

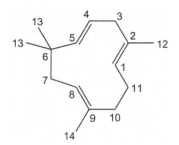

1. The molecular formula is $C_{15}H_{24}$. Calculate the degree of unsaturation from the molecular formula: the effective molecular formula is $C_{15}H_{24}$ (C_nH_m) which gives the degree of unsaturation as $(n - m/2 + 1) = 15 - 12 + 1 = 4$. The compound contains a combined total of four rings and/or π bonds.

2. From the question, we are told that the compound is a macrocycle (*i.e.* a ring compound) and contains three double bonds. The compound must contain only one ring to account for all of the degrees of unsaturation.

^{1}H NMR spectrum of α-humulene (CDCl₃, 600 MHz, 298 K)

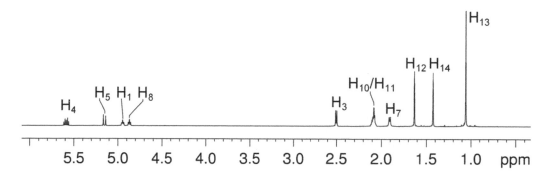

$^{13}C\{^1H\}$ NMR spectrum of α-humulene ($C_2D_2Cl_4$, 150 MHz, 338 K)

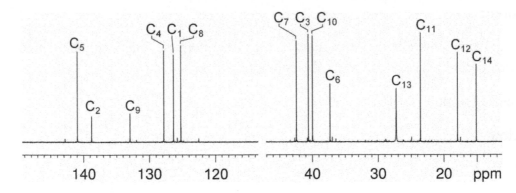

3. It is easiest to begin by determining the multiplicity of each ^{13}C resonance using integrations from the ^1H NMR spectrum and the me-^1H–^{13}C HSQC spectrum. We will give each carbon nucleus a temporary designation to identify it in the table below.

Signal	Temporary Designation	Multiplicity	Neighbours
140.9	C_A	CH	C_D, C_J
138.8	C_B	C	C_E, C_H, C_M
132.9	C_C	C	C_F, C_I, C_N
127.8	C_D	CH	C_A, C_H
126.3	C_E	CH	C_B, C_L
125.2	C_F	CH	C_C, C_G
42.4	C_G	CH_2	C_F, C_J, C_K
40.6	C_H	CH_2	C_B, C_D
39.9	C_I	CH_2	C_C, C_L
37.3	C_J	C	C_A, C_G, C_K
27.3	C_K	$2 \times CH_3$	C_J
23.6	C_L	CH_2	C_E, C_I
18.0	C_M	CH_3	C_B
15.1	C_N	CH_3	C_C

^1H–^{13}C me-HSQC spectrum of α-humulene ($C_2D_2Cl_4$, 600 MHz, 298 K)

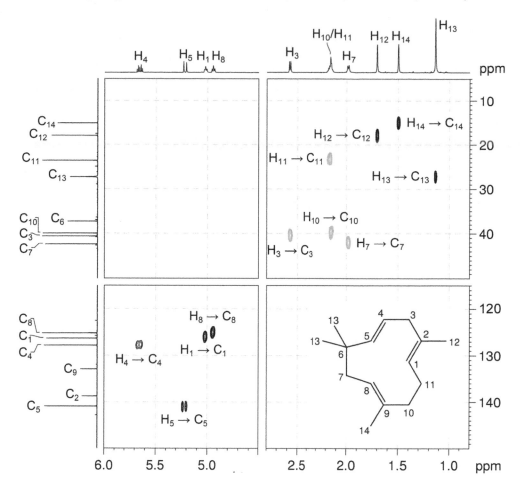

4. From the chemical shifts, we will assume that the low-field resonances in the ^1H and ^{13}C NMR spectra correspond to alkene groups.

5. We can now use the INADEQUATE spectrum to piece together the skeleton of the molecule.

6. Pick a starting point – in this case, we will start with C_A, the lowest-field resonance, which is a methine group (CH). There are two correlations for C_A – one to the signal at 37.3 ppm (C_J) and the other to the signal at 127.8 ppm (C_D). We know, then, that C_A is directly bond to C_J (a quaternary carbon centre) and C_D (a methine group (CH)):

INADEQUATE spectrum of α-humulene (C₂D₂Cl₄, 150 MHz, 338 K)

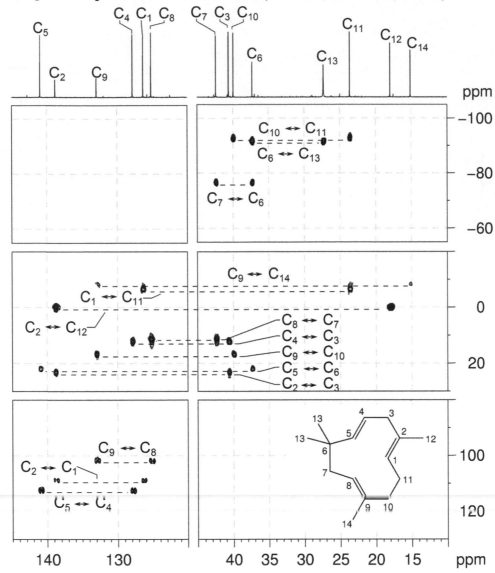

7. We will only work in one direction, so let us work from C_J. This resonance has two further correlations in the INADEQUATE spectrum – one to the resonance at 27.3 ppm (C_K, which, according to the integrations in the ¹H NMR spectrum, corresponds to a pair of geminal methyl groups) and the second to the resonance at 42.4 ppm (C_G, a methylene group).

$$C_D{=}C_A$$
H ⟍ ⟋H
H ⟋ $C_J{-}C_GH_2$
C_KH_3 C_KH_3

8. C_K has no further correlations as it is a pair of terminal geminal methyl groups.

9. C_G correlates to the signal at 125.2 ppm (C_F), a methine (CH) group.

$$C_D=C_A \quad (H) \quad C_J-C_GH_2-C_FH \quad C_KH_3 \quad C_KH_3$$

10. C_F also correlates to the resonance at 132.9 ppm (C_C, a quaternary carbon atom).

$$C_D=C_A \quad C_J-C_GH_2-C_F \quad C_C \quad C_KH_3 \quad C_KH_3 \quad H$$

11. C_C has two further correlations – the first to the resonance at 39.9 ppm (C_I, a methylene group) and the second to the resonance at 15.1 ppm (C_N, a methyl group).

$$C_D=C_A \quad C_NH_3 \quad C_C-C_IH_2 \quad C_J-C_GH_2-C_F \quad C_KH_3 \quad C_KH_3 \quad H$$

12. C_N has no further correlations as it is a terminal methyl group.

13. C_I also correlates to the signal at 23.6 (C_L, a methylene group).

$$C_D-C_A \quad C_NH_3 \quad C_C-C_IH_2-C_LH_2 \quad C_J-C_GH_2-C_F \quad C_KH_3 \quad C_KH_3 \quad H$$

14. C_L further corresponds to the resonance at 126.3 ppm (C_E, a methine carbon).

$$C_D=C_A \quad C_NH_3 \quad C_C-C_IH_2-C_LH_2-C_EH \quad C_J-C_GH_2-C_F \quad C_KH_3 \quad C_KH_3 \quad H$$

15. C_E also correlates to the signal at 138.8 (C_B, a quaternary alkene carbon).

$$C_D=C_A \quad C_NH_3 \quad C_B \quad C_C-C_IH_2-C_LH_2-C_E \quad C_J-C_GH_2-C_F \quad C_KH_3 \quad C_KH_3 \quad H \quad H$$

16. C$_B$ correlates to two additional resonances, the first is the signal at 40.6 ppm (C$_H$, a methylene group) and the second is the signal at 18.0 ppm (C$_M$, a methyl group).

17. C$_M$ has no further correlations as it is a terminal methyl group.

18. C$_H$ also correlates to the signal at 127.8 ppm (C$_D$), so we have closed the ring, and assigned all the correlations in the spectrum.

19. We can redraw the structure to give a more easily understood picture of the compound (showing all the double bonds as *trans*), and renumber to give the final answer:

Question:

Identify the following compound.

Molecular Formula: $C_{10}H_{10}O_3$

IR: 3300–2400 (br), 1685 cm^{-1}

Solution:

3,4-Dihydro-2H–benzopyran-3-carboxylic acid

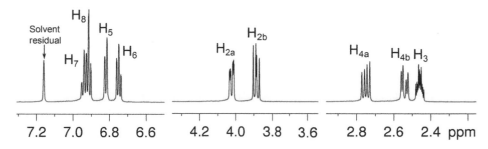

1. The molecular formula is $C_{10}H_{10}O_3$. Calculate the degree of unsaturation from the molecular formula: ignore the O atoms to give an effective molecular formula of $C_{10}H_{10}$ (C_nH_m) which gives the degree of unsaturation as $(n - m/2 + 1) = 10 - 5 + 1 = 6$. The compound contains a combined total of six rings and/or π bonds.

2. IR and ID NMR spectra establish the presence of an OH group (exchangeable proton at 11.95 ppm) and carbonyl group (^{13}C resonance at 179.0 ppm).

3. In the ^1H NMR spectrum, there are signals for four aromatic protons and the coupling pattern indicates an *ortho*-disubstituted aromatic ring where the signals at 6.91 and 6.82 ppm are due to the protons adjacent to the substituents (H_5 or H_8) whilst the signals at 6.94 and 6.75 ppm are due to the remaining aromatic protons (H_6 or H_7).

4. The aromatic ring and carbonyl group account for five degrees of unsaturation and the remaining degree of unsaturation is likely due to another ring in the molecule as there are no other signals for groups containing double bonds.

5. There are also five signals integrating for one proton each in the aliphatic region of the ^1H NMR spectrum.

^1H NMR spectrum of 3,4-dihydro-2H–benzopyran-3-carboxylic acid (C_6D_6, 600 MHz)

$^{13}C\{^1H\}$ NMR spectrum of 3,4-dihydro-2H–benzopyran-3-carboxylic acid (C$_6$D$_6$, 150 MHz)

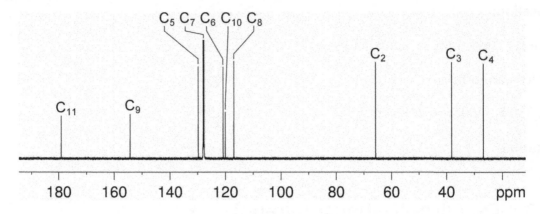

6. In the me-HSQC spectrum, there are two CH$_2$ groups with diastereotopic protons (^{1}H 4.02, 3.89 ppm / ^{13}C 65.7, H$_{2a}$,H$_{2b}$/C$_2$ and ^{1}H 2.75, 2.54 ppm / ^{13}C 26.8, H$_{4a}$,H$_{4b}$/C$_4$) and one CH group (^{1}H 2.46 ppm / ^{13}C 38.1, H$_3$/C$_3$).

7. C$_2$ is bound to oxygen on the basis of chemical shift.

^{1}H–^{13}C me-HSQC spectrum of 3,4-dihydro-2H–benzopyran-3-carboxylic acid (C$_6$D$_6$, 600 MHz)

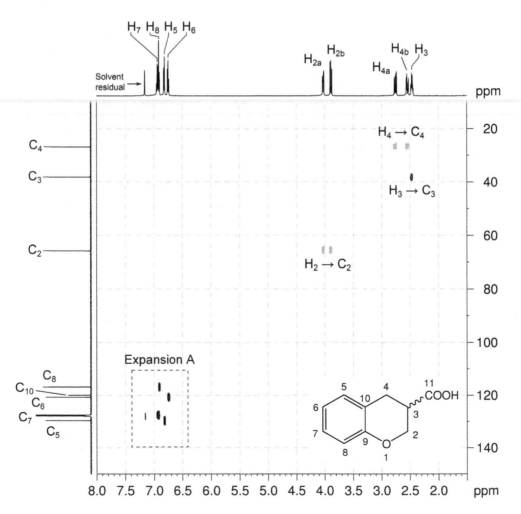

8. The protonated aromatic carbons C_5, C_7, C_6 and C_8 can be identified using the me-HSQC spectrum (expansion A) as ^{13}C signals at 129.7, 127.6, 120.8 and 116.9 ppm.

9. The ^{13}C signals for the quaternary aromatic signals are thus at 154.3 and 120.0 ppm.

^1H–^{13}C me-HSQC spectrum of 3,4-dihydro-2H–benzopyran-3-carboxylic acid – expansion A

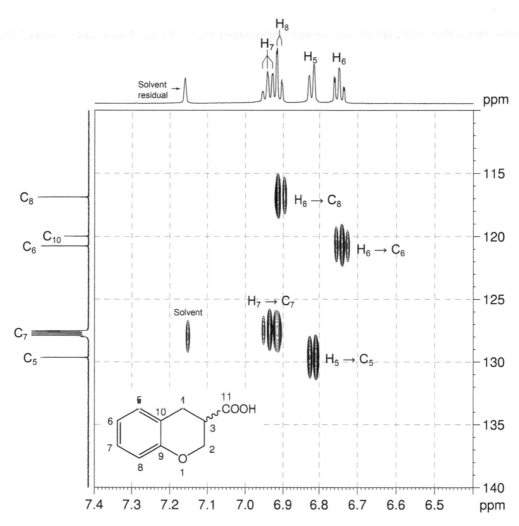

10. In the COSY spectrum, there are the expected aromatic proton correlations and also correlations from both sets of CH_2 protons to the CH proton. Based on the coupling pattern in the aliphatic region of the ^1H NMR spectrum, we can deduce the $-OCH_2CHCH_2-$ fragment. The H_2-C_3/C_4, H_3-C_2/C_4 and H_4-C_2/C_3 correlations in the HMBC spectrum are consistent with this fragment.

$^1H-^1H$ COSY spectrum of 3,4-dihydro-2H–benzopyran-3-carboxylic acid (C_6D_6, 600 MHz)

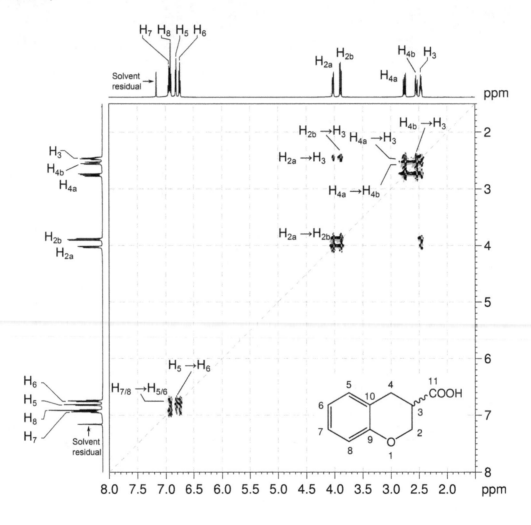

11. In the HMBC spectrum, the correlations of the aliphatic protons to the carbonyl carbon (H2–C11, H4–C11, H3–C11) show that the carbonyl carbon must be bound to C3 to afford the OCH2CH(C=O)CH2 fragment.

12. The correlation from the aromatic proton at 6.82 ppm to the CH2 carbon at 26.8 ppm in the HMBC spectrum (H5–C4) shows that the CH2 group is a substituent of the aromatic ring. This correlation also tells us that the ^{1}H signal at 6.82 ppm is due to the aromatic proton adjacent to the CH2 substituent.

^{1}H–^{13}C HMBC spectrum of 3,4-dihydro-2H–benzopyran-3-carboxylic acid (C$_6$D$_6$, 600 MHz)

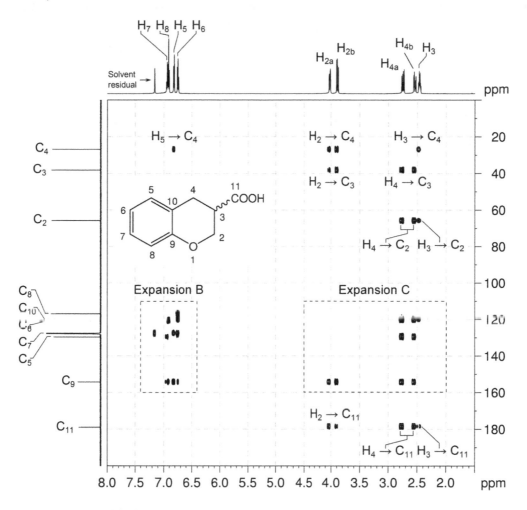

13. There are two possible isomers:

A   B

14. The ^{1}H chemical shift of the OH proton (11.95 ppm) points towards Isomer B as the correct compound, however the structure should be verified using the HMBC spectrum.

15. In the HMBC spectrum (expansion B), the H_5 proton correlates to ^{13}C signals at 127.6 and 154.3 ppm. The signal at 127.6 ppm is due to a protonated aromatic carbon and can be assigned to C_7. The signal at 154.3 ppm is due to a quaternary aromatic carbon and can be assigned to C_9.

16. Using the me-HSQC spectrum, H_7 can be assigned to the ^1H signal at 6.94 ppm.

17. Based on the coupling pattern, the ^1H signal at 6.91 ppm must be due to H_8 and the remaining ^1H signal at 6.75 ppm due to H_6. C_8 and C_6 can be assigned to ^{13}C signals at 116.9 and 120.8 ppm, respectively, using the me-HSQC spectrum.

18. In the HMBC spectrum (expansion B), the $H_7 \rightarrow C_5 / C_9$ as well as the $H_8 \rightarrow C_6 / C_9$ correlations are consistent with the structure. H_6 correlates to C_7, C_8, C_9 and the remaining quaternary ^{13}C signal at 120.0 ppm which must be due to C_{10}.

$^1H-^{13}C$ HMBC spectrum of 3,4-dihydro-2H–benzopyran-3-carboxylic acid – expansion B

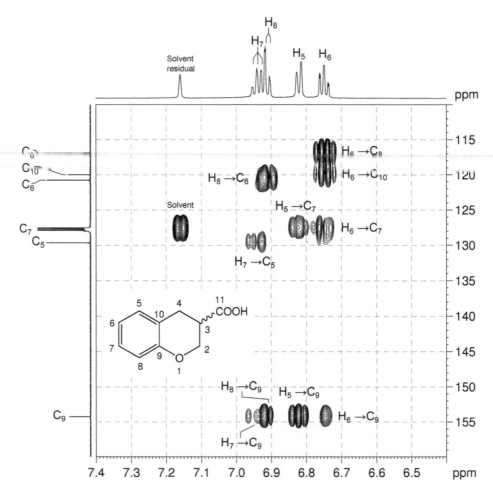

19. In expansion C of the HMBC spectrum, the $H_2 \rightarrow C_9$, $H_3 \rightarrow C_{10}$ and $H_4 \rightarrow C_5/C_9/C_{10}$ correlations identify the correct structure as that of **isomer B**. In particular note that the $H_2 \rightarrow C_9$ correlation would be an unlikely five-bond coupling in isomer A.

^1H–^{13}C HMBC spectrum of 3,4-dihydro-2H–benzopyran-3-carboxylic acid – expansion C

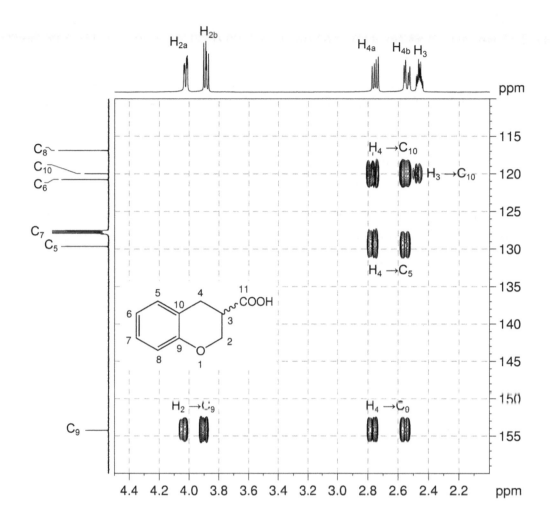

Problem 58

Question:

The ^1H and $^{13}C\{^1H\}$ NMR spectra of quinidine ($C_{20}H_{24}N_2O_2$) recorded in DMSO-d_6 solution at 298 K and 600 MHz are given below.

The ^1H NMR spectrum has signals at δ 1.38 (m, 1H), 1.43 (m, 2H), 1.68 (m, 1H), 1.94 (m, 1H), 2.17 (m, 1H), 2.53 (m, 1H), 2.63 (m, 1H), 2.70 (m, 1H), 3.00 (m, 1H), 3.06 (m, 1H), 3.90 (s, 3H), 5.05 (m, 1H), 5.08 (m, 1H), 5.32 (dd, J = 4.7, 6.2 Hz, 1H), 5.73 (d, J = 4.7 Hz, 1H), 6.10 (ddd, J = 7.5, 10.2, 17.5 Hz, 1H), 7.39 (dd, J = 2.8, 9.3 Hz, 1H), 7.48 (d, J = 2.8 Hz, 1H), 7.53 (d, J = 4.4 Hz, 1H), 7.94 (d, J = 9.3 Hz, 1H) and 8.69 (d, J = 4.4 Hz, 1H) ppm.

The $^{13}C\{^1H\}$ NMR spectrum has signals at δ 23.1, 26.2, 27.8, 39.8, 48.4, 49.1, 55.3, 60.5, 70.7, 102.3, 114.2, 118.9, 120.9, 126.9, 131.0, 141.2, 143.8, 147.4, 149.4 and 156.7 ppm.

The 2D ^1H–^1H COSY, multiplicity-edited ^1H–^{13}C HSQC, ^1H–^{13}C HMBC and ^1H–^1H NOESY spectra are given on the following pages. Use these spectra to assign the ^1H and $^{13}C\{^1H\}$ resonances for this compound.

Solution:

Quinidine

Proton	Chemical Shift (ppm)	Carbon	Chemical Shift (ppm)
H_1	8.69	C_1	147.4
H_2	7.53	C_2	118.9
		C_3	149.4
		C_4	126.9
H_5	7.48	C_5	102.3
		C_6	156.7
H_7	7.39	C_7	120.9
H_8	7.94	C_8	131.0
		C_9	143.8
H_{10}	5.32	C_{10}	70.7
H_{11}	3.00	C_{11}	60.5
H_{12a}	1.94	C_{12}	23.1
H_{12b}	1.38		
H_{13}	1.68	C_{13}	27.8
H_{14a}	1.43	C_{14}	26.2
H_{14b}	1.43		
H_{15a}	2.53	C_{15}	49.1
H_{15b}	2.63		
H_{16a}	2.70	C_{16}	48.4
H_{16b}	3.06		
H_{17}	2.17	C_{17}	39.8
H_{18}	6.10	C_{18}	141.2
H_{19a}	5.08	C_{19}	114.2
H_{19b}	5.05		
H_{20}	3.90	C_{20}	55.3
OH	5.73		

1. The aromatic protons can be assigned from the coupling patterns and constants in the ^1H NMR spectrum. Based on the coupling constant of 2.8 Hz, the signal at 7.48 ppm is assigned to H_5, which has no neighbouring protons in *ortho* positions.

2. The signal at 7.39 ppm (dd, $J = 2.8$ and 9.3 Hz) is assigned to H_7 (coupling to H_5 ($J = 2.8$ Hz) and its *ortho* neighbour H_8 ($J = 9.3$ Hz).

3. The signal at 7.94 ppm is a doublet with a splitting of 9.3 Hz, so this is assigned to H_8.

^{1}H NMR spectrum of quinidine – expansion (DMSO-d_6, 600 MHz)

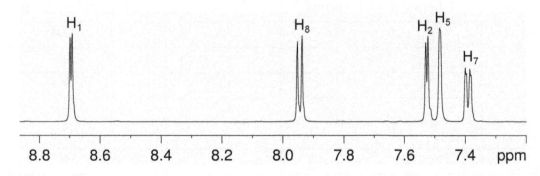

4. The remaining two signals, which are mutually-coupled doublets with $J = 4.4$ Hz, are assigned to H_1 and H_2. Based on the chemical shift, the signal at 8.69 ppm is assigned as H_1 (as it is adjacent to the nitrogen atom) and the signal at 7.53 ppm is assigned to H_2.

5. On the basis of integration, the signal at 3.90 ppm is assigned to H_{20}.

6. The me-HSQC spectrum (expansion F) is used to assign the protonated carbon atoms of the aromatic ring: C_1 at 147.4, C_2 at 118.9, C_5 at 102.3, C_7 at 120.9 and C_8 at 131.0 ppm.

7. This spectrum also allows identification of the protonated alkene carbon atoms: C_{19} – the methylene carbon is at 114.3 ppm. H_{19a} and H_{19b} are identified as the signals at 5.08 and 5.05 ppm in the ^1H NMR spectrum. C_{18} is the remaining protonated carbon resonance at 141.2 ppm, and H_{18} is the ^1H NMR resonance at 6.10 ppm.

$^1H–^{13}C$ me-HSQC spectrum of quinidine – expansion F (DMSO-d_6, 600 MHz)

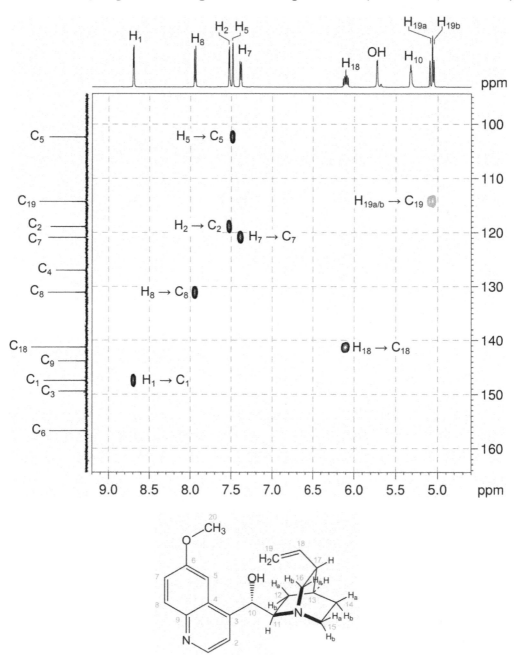

$^1H–^{13}C$ me-HSQC spectrum of quinidine – expansion E

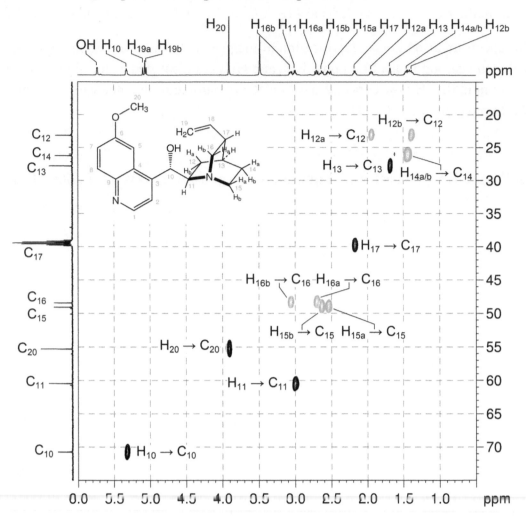

8. C_{20} is assigned as the signal at 55.3 ppm.

9. Expansion E of the me-HSQC spectrum can be used to identify which resonances in the alkyl region of the spectrum correspond to methine (CH) resonances, and which correspond to diastereotopic methylene (CH_2) groups. There are resonances for four CH groups at ^1H 5.32 / ^{13}C 70.7 ppm (H_{10} / C_{10}), ^1H 3.00 / ^{13}C 60.5 ppm (H_{11} / C_{11}), ^1H 2.17 / ^{13}C 39.8 ppm (H_{17} / C_{17}) and ^1H 1.68 / ^{13}C 27.8 ppm (H_{13} / C_{13}). There are four CH_2 groups, all with diastereotopic protons, at ^1H 3.06, 2.70 / ^{13}C 48.4 ppm (H_{16a}, H_{16b} / C_{16}), ^1H 2.63, 2.53 / ^{13}C 49.1 ppm (H_{15a}, H_{15b} / C_{15}), ^1H 1.94,1.38 / ^{13}C 23.1 ppm (H_{12a}, H_{12b} / C_{12}) and ^1H 1.43,1.43 / ^{13}C 26.2 ppm (H_{14a}, H_{14b} / C_{14}).

10. Close inspection of the HSQC spectrum shows that the ^1H resonance at 5.73 ppm does not correlate to a carbon resonance. This resonance must therefore correspond to the hydroxyl proton.

11. The ^{1}H–^{1}H COSY spectrum can now be used to identify the resonances in the quinuclidine portion of the molecule.

^{1}H–^{1}H COSY spectrum of quinidine – expansion A (DMSO-d_6, 600 MHz)

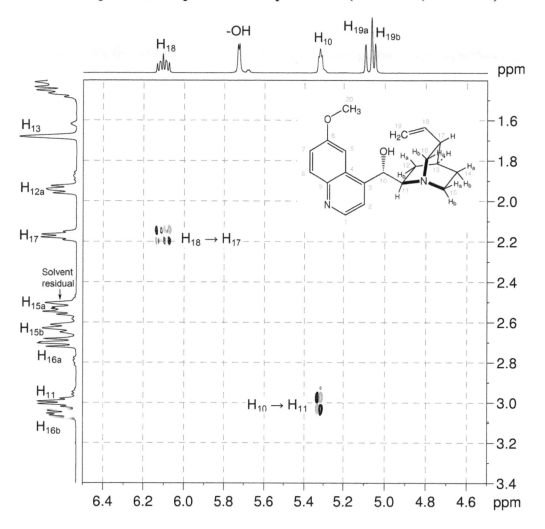

12. Expansion A shows a correlation between H_{18} and the signal at 2.17 ppm. This is assigned as H_{17}.

¹H–¹H COSY spectrum of quinidine – expansion C

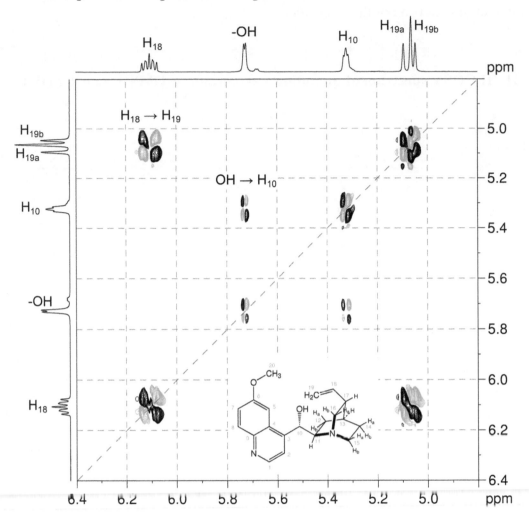

13. Expansion C shows the expected correlation between H_{18} and H_{19a} and H_{19b}, and a second correlation between the hydroxyl proton and the signal at 5.32 ppm. This resonance is assigned as H_{10} (this resonance could also have been assigned on the basis of chemical shift).

14. In expansion A, H_{10} correlates to the signal at 3.00 ppm, which is assigned as H_{11}.

15. In expansion B, H_{11} correlates to the diastereotopic methylene pair at 1.94 and 1.38 ppm. These signals are assigned as H_{12a} and H_{12b}, but we cannot yet tell which is which.

^1H–^1H COSY spectrum of quinidine – expansion B

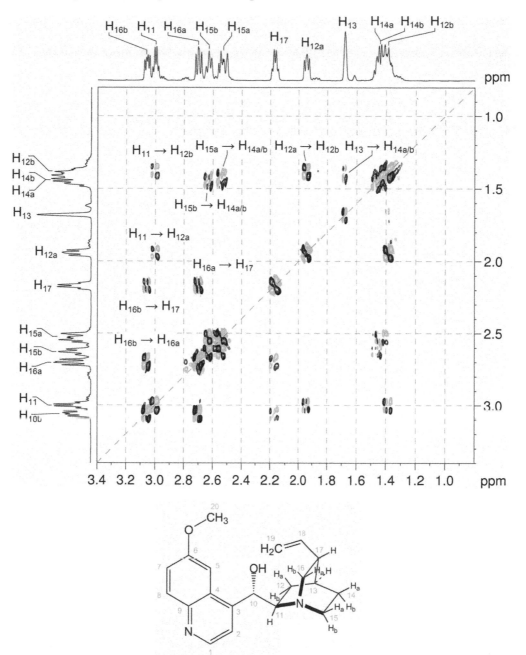

16. There is also a correlation between H_{12a} and H_{12b} in expansion B. These resonances, which should correlate to H_{13}, show no additional cross-peaks. The bond angle between H_{13} and the neighbouring diastereotopic H_{12} protons must render the coupling vanishingly small (see Karplus relationship). Nevertheless, H_{13} can be identified as the only remaining unassigned CH signal – at 1.68 ppm.

17. H_{13} correlates to the methylene signals at 1.43 ppm, which is assigned as H_{14a} and H_{14b} (these signals are virtually coincident, and we cannot differentiate them). In turn, this methylene group correlates to the methylene signals at 2.63 and 2.53 ppm, which are assigned as H_{15a} and H_{15b} – once again, we cannot yet tell which resonance belongs to which proton.

18. The remaining diastereotopic pair – at 3.06 and 2.70 ppm – is assigned as H_{16a} and H_{16b}. Both show the expected correlations to each other, as well as H_{17}. We cannot yet tell which resonance belongs to which proton.

^{1}H–^{1}H COSY spectrum of quinidine – expansion D

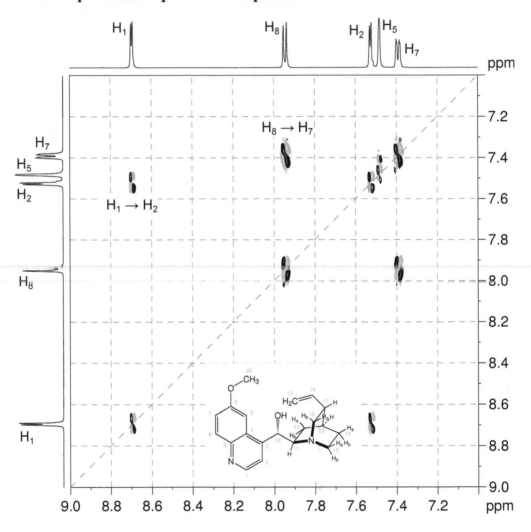

19. Each protonated carbon can now be identified using the me-HSQC spectrum.

20. The HMBC spectrum is used to identify the non-protonated carbon atoms. **<u>Remember</u>** that: a) in aromatic systems, the three-bond coupling $^3J_{C-H}$ is *typically* the larger long-range coupling and gives rise to the strongest cross-peaks; b) the presence of electronegative substituents, such as oxygen or nitrogen, *may* increase the size of $^2J_{C-H}$; c) benzylic protons correlate to the *ipso* carbon (two-bond correlation) and the *ortho* carbons (three-bond correlations).

$^1H–^{13}C$ HMBC spectrum of quinidine – expansion G (DMSO-d_6, 600 MHz)

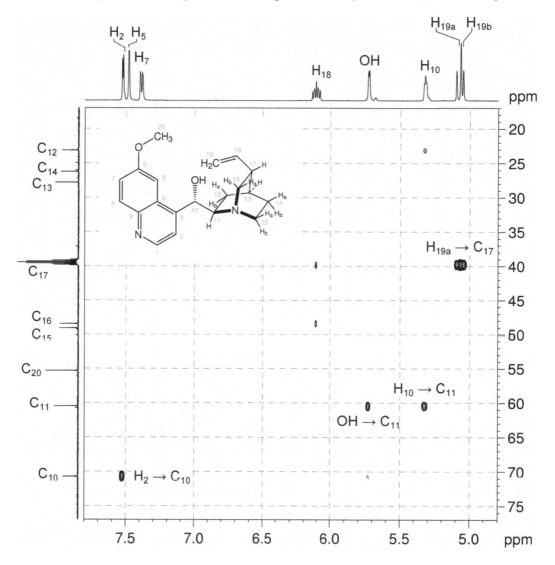

^{1}H–^{13}C HMBC spectrum of quinidine – expansion H

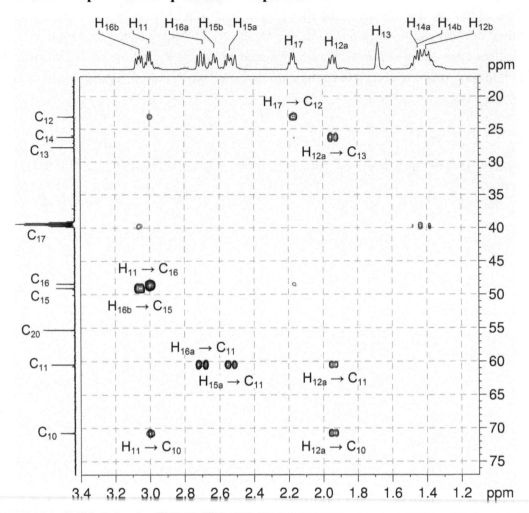

21. H_{20} correlates to the signal at 156.7 ppm (expansion J). This resonance is assigned as C_6.

^{1}H–^{13}C HMBC spectrum of quinidine – expansion J

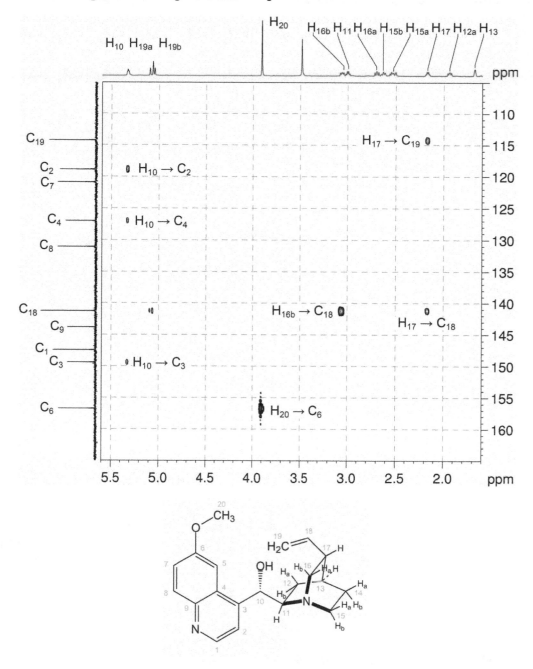

22. H$_7$ correlates to the resonances at 102.3 ppm (C$_5$) and 143.8 ppm (expansion I). This latter resonance is assigned to C$_9$.

23. H$_8$ correlates to the resonances at 126.9 ppm and 156.7 ppm (C$_6$) (expansion I). The former resonance is assigned to C$_4$.

24. H$_1$ correlates to the resonances at 118.9 ppm (C$_2$), 143.8 ppm (C$_9$) and 149.4 ppm. The latter resonance is assigned to C$_3$.

^{1}H–^{13}C HMBC spectrum of quinidine – expansion I

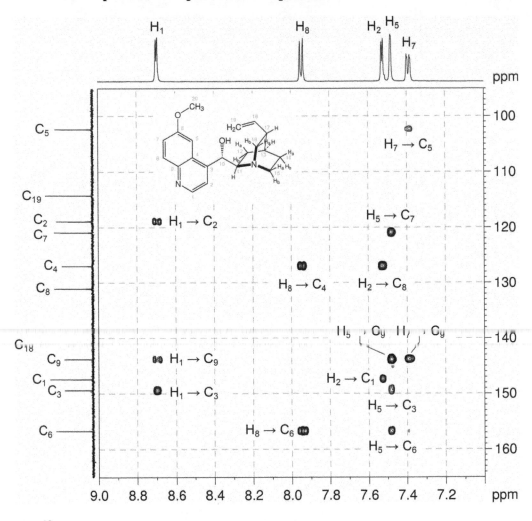

25. All the ^{13}C resonances have now been assigned.

^{13}C{^{1}H} NMR spectrum of quinidine (DMSO-d_6, 150 MHz)

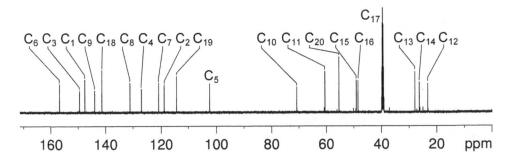

26. The NOESY spectrum can be used to identify which diastereotopic protons occur on which face of the quinuclidine ring system.

27. H$_{11}$ shows a NOE correlation to the signal at 1.38 ppm (expansion K), previously identified as either H$_{12a}$ or H$_{12b}$. As H$_{11}$ is on the bottom face of the quinuclidine ring, we can now assign the signal at 1.38 ppm as H$_{12b}$. The resonance at 1.94 ppm is therefore assigned as H$_{12a}$. This assignment is confirmed by the correlation between the signal at 1.94 ppm and H$_{18}$ (expansion L).

^{1}H–^{1}H NOESY spectrum of quinidine – expansion K (DMSO-d_6, 600 MHz)

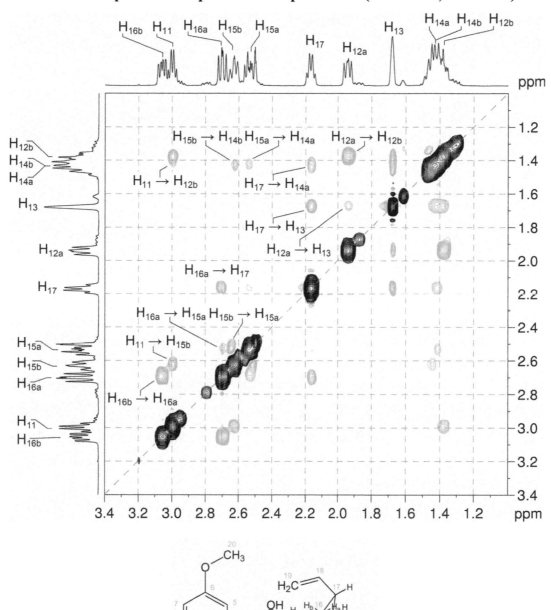

$^1H–^1H$ NOESY spectrum of quinidine – expansion L

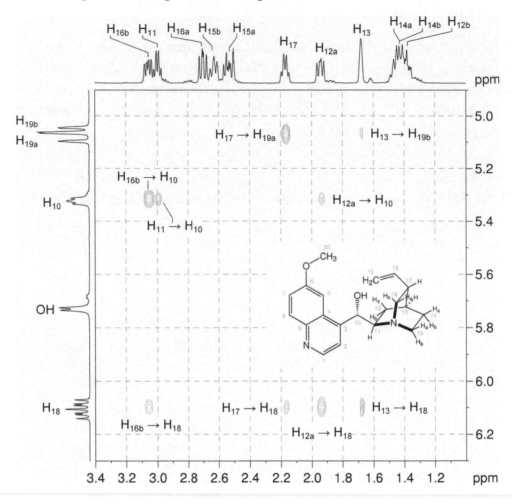

28. H_{11} also correlates to the signal at 2.63 ppm, previously identified as either H_{15a} or H_{15b} (expansion K). As H_{11} is on the bottom face of the quinuclidine ring, we can now assign the signal at 2.63 ppm as H_{15b}. The resonance at 2.53 ppm is therefore assigned as H_{15a}.

29. H_{17} correlates to the signal at 2.70 ppm (expansion K), previously identified as either H_{16a} or H_{16b}. We can now assign this signal as H_{16a}. The resonance at 3.06 ppm is therefore assigned as H_{16b}. This assignment is confirmed by the correlation between the signal at 3.06 ppm and H_{18} (expansion L).

30. The alkene protons H_{19a} and H_{19b} may be assigned based on coupling constants. The ^1H signal at 5.08 ppm has a coupling constant of 17 Hz and can be assigned to the proton *trans* to H_{18} (H_{19a}). The signal at 5.05 ppm has a coupling constant of 10 Hz and can be assigned to the proton *cis* to H_{18} (H_{19b}).

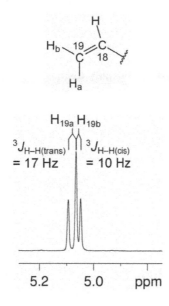

^1H NMR spectrum of quinidine – expansions (DMSO-d_6, 600 MHz)

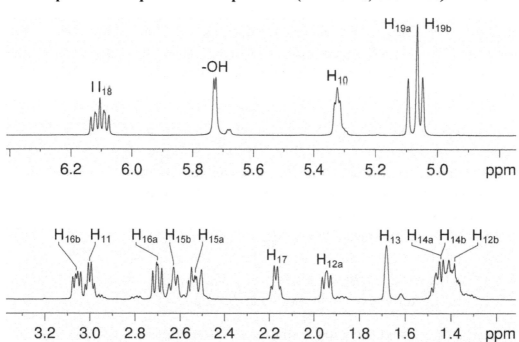

Problem 59

Question:

Identify the following compound.

Molecular Formula: $C_{13}H_{21}NO_3$

HINT: This compound contains a secondary amine.

Solution:

Salbutamol

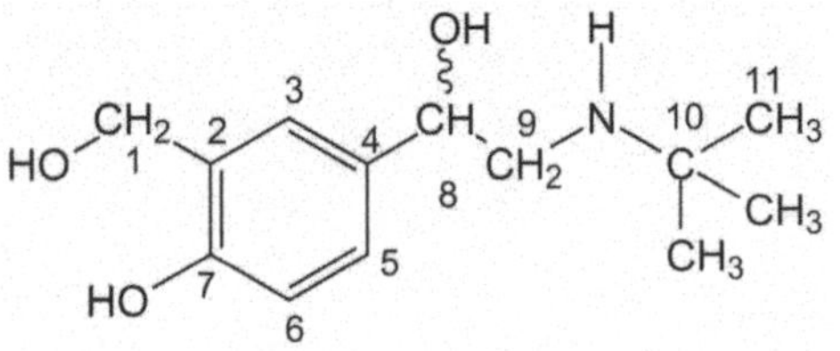

1. The molecular formula is $C_{13}H_{21}NO_3$. Calculate the degree of unsaturation from the molecular formula: ignore the O atoms and ignore the N and remove one H to give an effective molecular formula of $C_{13}H_{20}$ (C_nH_m) which gives the degree of unsaturation as $(n - m/2 + 1) = 13 - 10 + 1 = 4$. The compound contains a combined total of four rings and/or π bonds.

2. There are four exchangeable protons – hint given that the compound contains a secondary amine *i.e.* there is an NH group present, thus the remaining three exchangeable protons must be OH groups. The three hydroxy groups account for all the oxygen atoms in the molecule.

3. From the ^1H NMR spectrum, there are three aromatic protons and the coupling pattern indicates the aromatic ring is a "1,2,4-trisubstituted benzene". The proton at 7.26 ppm clearly has no *ortho* couplings and must be the isolated proton in the spin system sandwiched between two substituents. The two protons at 6.99 and 6.69 ppm each have a large *ortho* coupling so they must be adjacent to each other. The proton at 6.99 ppm has an additional *meta* coupling so it must be the proton *meta* to the proton at 7.26 ppm. The remaining proton at 6.69 ppm must be the proton *para* to the proton at 7.26 ppm.

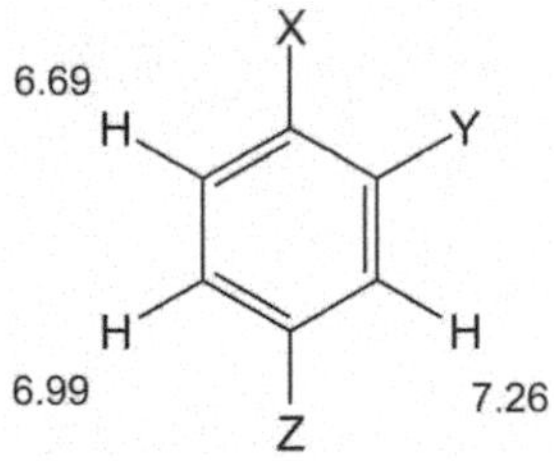

4. An aromatic ring would account for all degrees of unsaturation and there are no additional double bonds or rings in the structure.

5. Also from the ^1H NMR spectrum, there are signals for an isolated OCH_2 group at 4.46 ppm, an OCH group at 4.39 ppm, a CH_2 group at 2.53 ppm and three equivalent CH_3 groups at 1.01 ppm. The multiplicities of the signals can be confirmed using the $^1H-^{13}C$ me-HSQC spectrum.

^{1}H NMR spectrum of salbutamol (DMSO-d_6, 500 MHz)

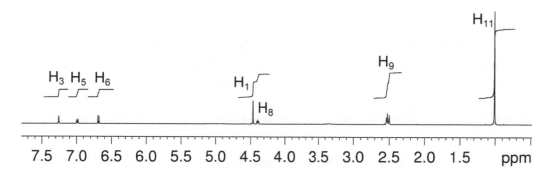

^{1}H NMR spectrum of salbutamol – expansions

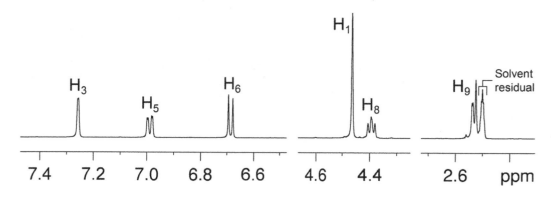

^{13}C{^{1}H} NMR spectrum of salbutamol (DMSO-d_6, 125 MHz)

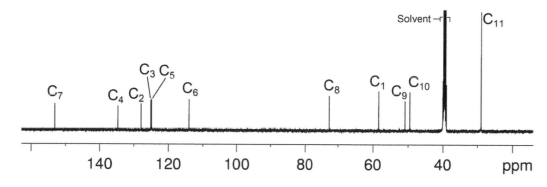

6. In the COSY spectrum, there are correlations between the OCH group at 4.39 ppm and the CH_2 group at 2.53 ppm indicating an OCH–CH_2 fragment.

7. In the COSY spectrum, there is also a weak correlation between the OCH_2 group at 4.46 ppm and the aromatic proton at 7.26 ppm indicating that the OCH_2 group is a substituent of the aromatic ring, most likely in a position *ortho* to the aromatic proton at 7.26 ppm (*i.e.* either at position Y or Z in the aromatic ring).

8. There are also the expected aromatic proton correlations in the COSY spectrum.

^1H–^1H COSY spectrum of salbutamol (DMSO-d_6, 500 MHz)

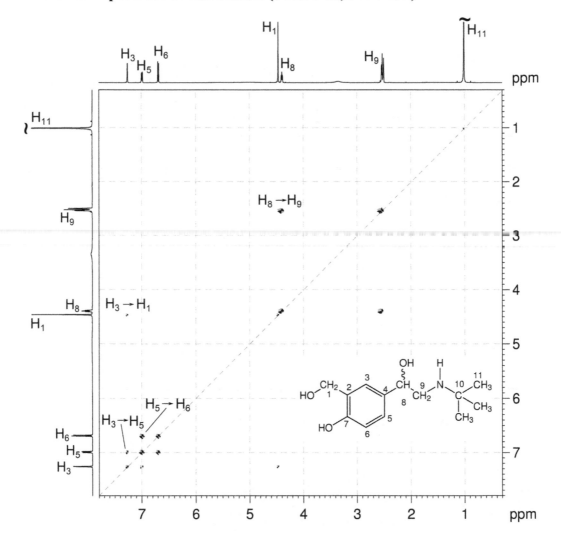

9. The ^1H–^{13}C me-HSQC spectrum easily identifies the protonated carbons. The isolated aromatic proton at 7.26 ppm correlates to

the carbon at 125.3 ppm. The aromatic protons at 6.99 and 6.69 ppm correlate to the carbons at 124.9 and 114.0 ppm, respectively.

10. The OCH proton at 4.39 ppm correlates to the carbon at 72.5 ppm. The OCH_2 protons at 4.46 ppm correlate to the carbon at 58.4 ppm. The CH_2 protons at 2.53 ppm correlate to the carbon at 50.9 ppm and the CH_3 protons correlate to the carbon at 29.0 ppm.

11. The three aromatic carbons which bear substituents appear at 153.1, 134.7 and 127.9 ppm. The low-field carbon resonance (153.1 ppm) must be a carbon bound to an oxygen atom.

^1H–^{13}C me-HSQC spectrum of salbutamol (DMSO-d_6, 500 MHz)

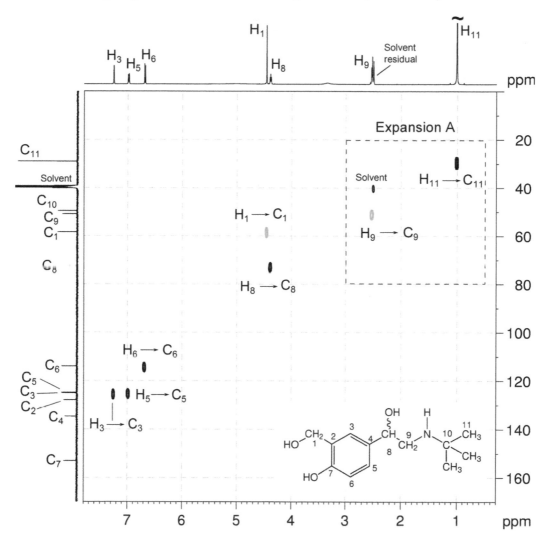

12. In the ^1H–^{13}C HMBC spectrum, the correlations from the aromatic protons at 7.26 and 6.99 ppm to the OCH carbon at 72.5 ppm place the OCH group as a substituent of the aromatic

ring adjacent to both protons at 7.26 and 6.99 ppm (*i.e.* in position Z of the ring).

^{1}H–^{13}C HMBC spectrum of salbutamol (DMSO-d_6, 500 MHz)

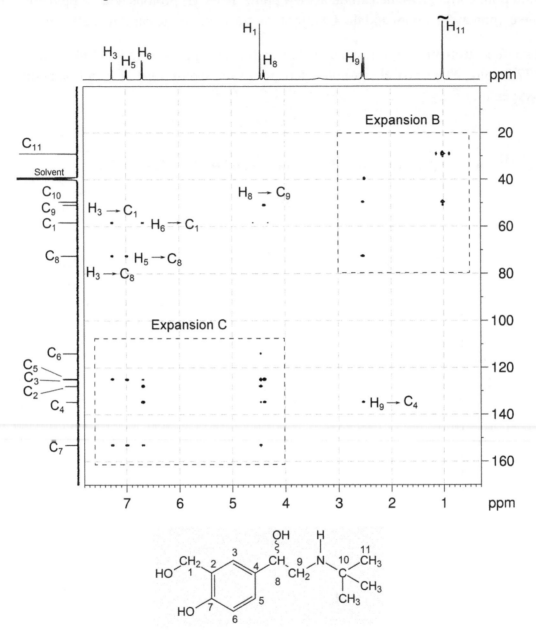

13. In expansion B of the ^{1}H–^{13}C HMBC spectrum, the methyl proton at 1.01 ppm correlates to the quaternary carbon at 49.5 ppm indicates the presence of a *tert*-butyl group.

14. Note that there is also a correlation from the methyl proton at 1.01 ppm to the carbon at 29.0 ppm. While this appears to be a one-bond correlation, in *tert*-butyl groups, isopropyl groups or compounds with a *gem*-dimethyl group, the apparent one-bond correlation arises from the $^3J_{C-H}$ interaction of the protons of one of the methyl groups with the chemically equivalent carbon which is three bonds away.

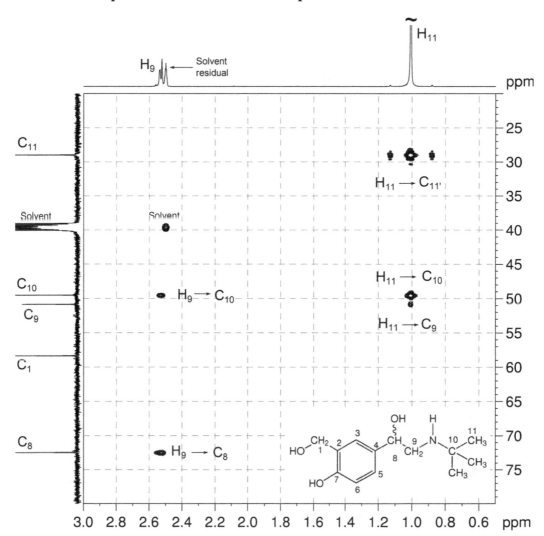

^{1}H–^{13}C HMBC spectrum of salbutamol – expansion B

15. As all oxygen atoms must be part of OH groups, we now have the following fragments:

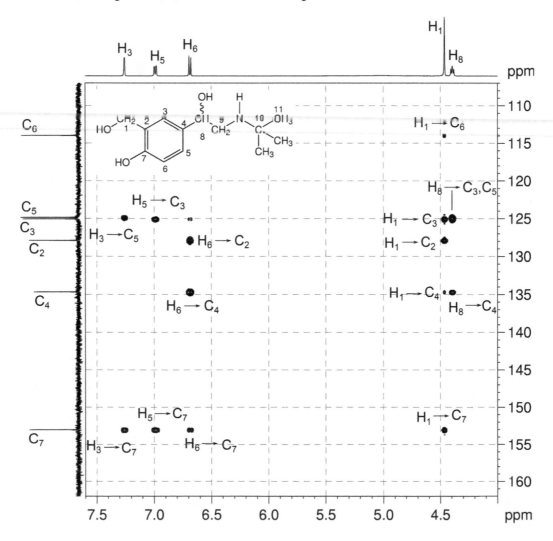

16. In expansion C of the HMBC spectrum, the aromatic proton at 7.26 ppm correlates to the protonated aromatic carbon at 124.9 ppm and the low-field quaternary carbon at 153.1 ppm. Therefore the low-field carbon resonance (bearing the –OH substituent) must be *meta* to the proton at 7.26 ppm (*i.e.* at position X in the aromatic ring).

^{1}H–^{13}C HMBC spectrum of salbutamol – expansion C

17. By elimination the OCH_2 group at 4.46 ppm must be in position Y in the ring and this is confirmed by the strong correlations from the OCH_2 proton at 4.46 ppm to the *ortho* carbons at 153.1 and 125.3 ppm. The remaining strong correlation to the carbon at 127.9 ppm identifies this resonance as the aromatic carbon directly bound to the OCH_2 group.

18. By elimination the quaternary aromatic carbon resonance at 134.7 ppm is due to the carbon bearing the OCH group (substituent Z).

19. We are now left with the following fragments:

20. The correlations in the NOESY spectrum are consistent with the structure. In particular, the correlations from the aromatic protons at 7.26 and 6.99 ppm to the OCH and CH_2 protons at 4.39 and 2.53 ppm confirm the placement of the $OCHCH_2$ fragment between these two aromatic protons in the ring (*i.e.* position Z).

21. Similarly the correlation from the aromatic proton at 7.26 ppm to the OCH_2 proton at 4.46 ppm places the OCH_2 fragment in the position adjacent to the proton at 7.26 ppm (*i.e.* position Y).

^1H–^1H NOESY spectrum of salbutamol (DMSO-d_6, 500 MHz)

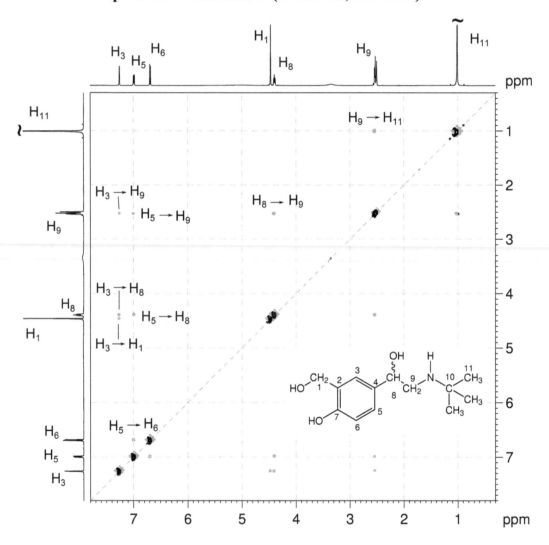

22. The ^1H–^{15}N HMBC spectrum shows two- and/or three-bond correlations between proton and nitrogen nuclei.

23. The NH group is located using the ^1H–^{15}N HMBC spectrum. The correlations from the CH$_2$ group at 4.39 ppm and the methyl group at 1.01 ppm to the N position the NH group between the CH$_2$ and *tert*-butyl groups.

^1H–^{15}N HMBC spectrum of salbutamol (DMSO-d_6, 500 MHz)

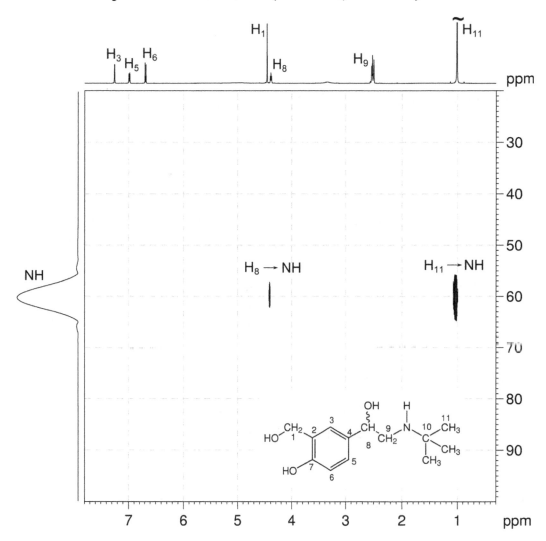

Question:

Identify the following compound.

Molecular Formula: $C_{11}H_8O_2$

IR: 2957, 2925, 2855, 1645, 1631, 1621 cm^{-1}.

Solution:

2-Hydroxy-1-naphthaldehyde

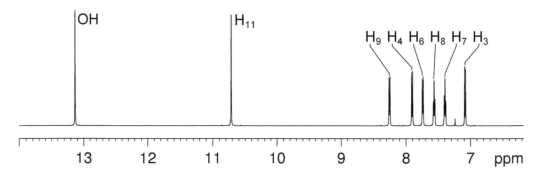

1. The molecular formula is $C_{11}H_8O_2$. Calculate the degree of unsaturation from the molecular formula: ignore the O atoms to give an effective molecular formula of $C_{11}H_8$ (C_nH_m) which gives the degree of unsaturation as $(n - m/2 + 1) = 11 - 4 + 1 = 8$. The compound contains a combined total of eight rings and/or π bonds.

2. From the ^1H NMR spectrum, there are two signals to low-field. The signal at 13.1 ppm is exchangeable and must be due to an –OH group.

3. There are also six aromatic signals which imply the presence of two aromatic rings.

4. The fact that there are only eleven carbons in the molecule indicates that the aromatic rings must be fused and share some common carbons.

^1H NMR spectrum of 2-hydroxy-1-naphthaldehyde (CDCl₃, 600 MHz)

$^{13}C\{^1H\}$ NMR spectrum of 2-hydroxy-1-naphthaldehyde (CDCl$_3$, 150 MHz)

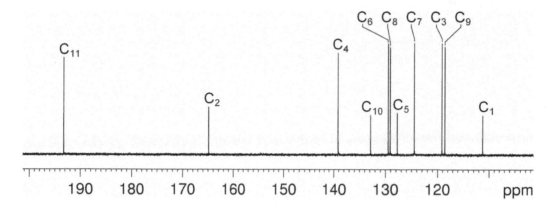

5. The ^1H–^1H COSY spectrum shows that there are two unique aromatic spin systems – one is a four spin system, the other a two spin system. From the magnitude of the coupling constants, we can conclude that the compound is either a 1,2- or 1,4-disubstituted naphthalene.

^1H–^1H COSY spectrum of 2-hydroxy-1-naphthaldehyde (CDCl$_3$, 600 MHz)

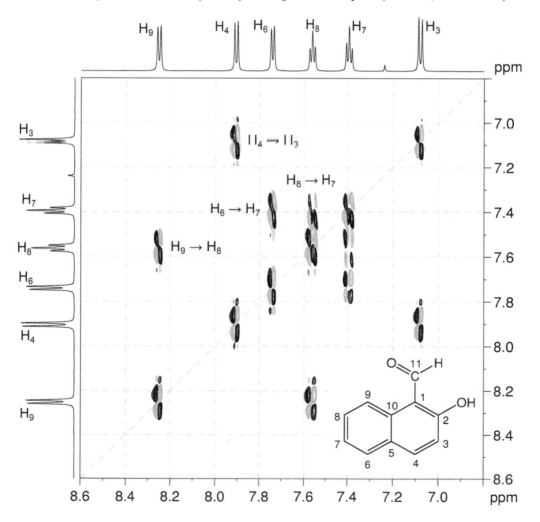

6. The ^1H–^{13}C me-HSQC spectrum shows that the ^1H signal at 10.7 ppm correlates to the ^{13}C signal at 193.4 ppm. This is consistent with the presence of an aldehyde.

7. All oxygen atoms in the molecular formula are now accounted for.

^1H–^{13}C me-HSQC spectrum of 2-hydroxy-1-naphthaldehyde (CDCl₃, 600 MHz)

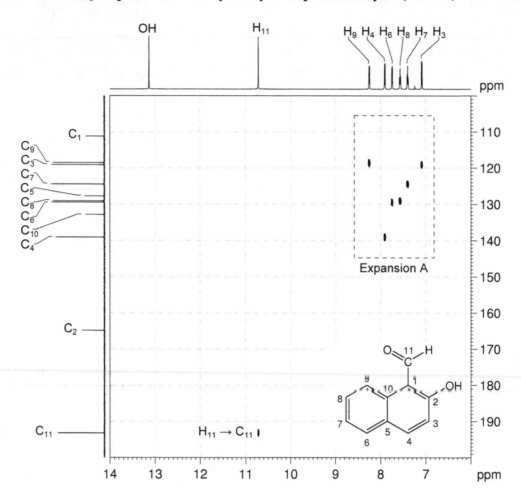

8. The ^{1}H–^{13}C me-HSQC spectrum can also be used to identify the protonated aromatic carbon atoms (at 139.1, 129.4, 129.1, 124.4, 119.1 and 118.5 ppm).

^{1}H–^{13}C me-HSQC spectrum of 2-hydroxy-1-naphthaldehyde – expansion A

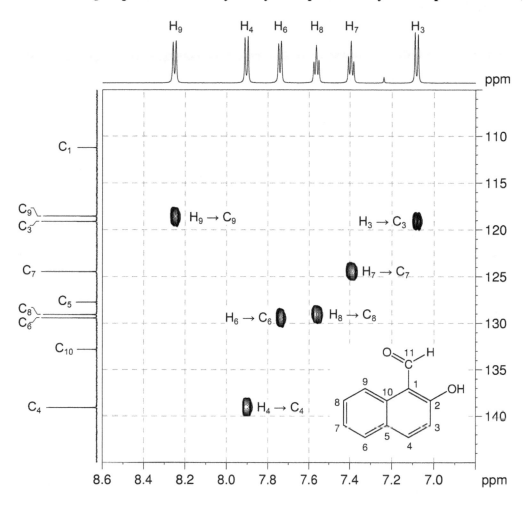

9. Based on the information available so far, there are three possible candidates:

10. The HMBC spectrum can be used to identify the correct isomer. **Remember** that: a) in aromatic systems, the three-bond coupling $^3J_{C-H}$ is *typically* the larger long-range coupling and gives rise to the strongest cross-peaks; b) the presence of electronegative substituents, such as oxygen or nitrogen, *may* increase the size of $^2J_{C-H}$; c) benzylic protons correlate to the *ipso* carbon (two-bond correlation) and the *ortho* carbons (three-bond correlations).

^{1}H–^{13}C HMBC spectrum of 2-hydroxy-1-naphthaldehyde (CDCl$_3$, 600 MHz)

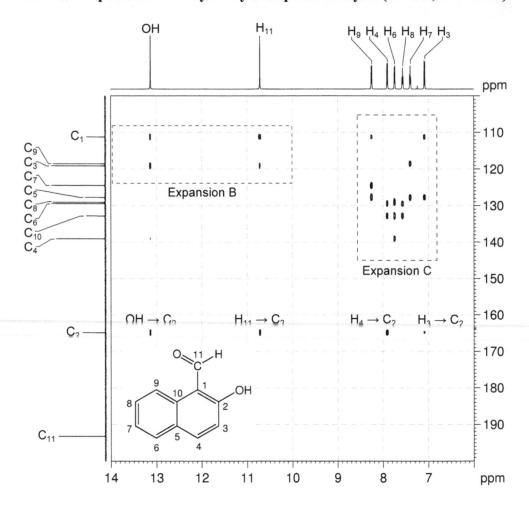

11. On the basis of chemical shift, the carbon atom *ipso* to the phenol group can be identified as the resonance at 164.9 ppm.

12. In expansion B, both the phenolic proton and the aldehydic proton correlate to the protonated carbon signal at 119.1 ppm, and the quaternary carbon signal at 111.2 ppm. The quaternary signal is assigned to the carbon atom *ipso* to the aldehyde (this can be confirmed via the observation that the $H_{11} \rightarrow C_1$ signal is split in expansion B, consistent with the large $^2J_{C-H}$ observed for aldehydes).

$^1H-^{13}C$ HMBC spectrum of 2-hydroxy-1-naphthaldehyde – expansion B

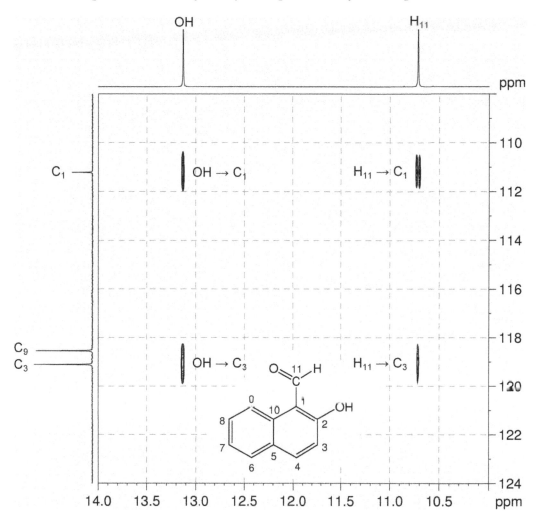

13. Isomer C can be ruled out, as the phenolic proton in this compound would not correlate to the carbon atom bearing the aldehyde (this would be a five-bond correlation).

14. From the $^1H-^1H$ COSY spectrum, we know that the four-spin aromatic ring has signals at 8.25, 7.74, 7.57 and 7.40 ppm. The first two signals represent the two protons on either end of the spin system, the second two signals the middle two protons.

15. In the ^1H–^{13}C HMBC (expansion C), there is a correlation between the signal at 8.25 ppm and the *ipso* carbon of the aldehyde (at 111.2 ppm). On this basis, we can eliminate Isomer B – as this would represent a four-bond correlation, while in Isomer A this is a three-bond correlation.

^1H–^{13}C HMBC spectrum of 2-hydroxy-1-naphthaldehyde – expansion C

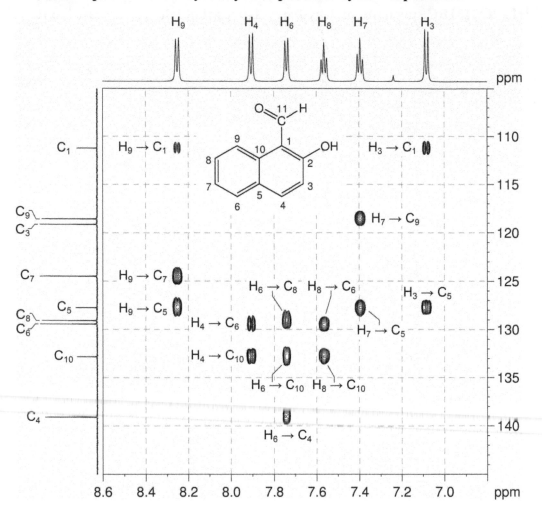

16. **Isomer A** (2-hydroxy-1-naphthaldehyde) is therefore the correct structure.

17. Note that the expected three-bond correlation between H_{11} and C_{10} is absent, and there is a four-bond correlation between H_{11} and C_3. A careful analysis of all the HMBC correlations supports the formulation of Isomer A over Isomer B.

Question:

Identify the following compound.

Molecular Formula: $C_{10}H_{10}O_2$

IR: 1689 cm^{-1}

Solution:

6-Methyl-4-chromanone

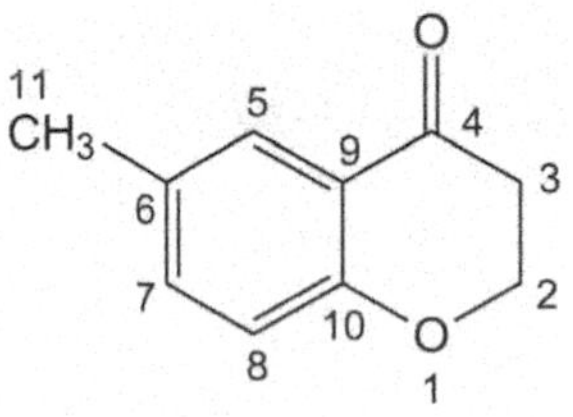

1. The molecular formula is $C_{10}H_{10}O_2$. Calculate the degree of unsaturation from the molecular formula: ignore the O atoms to give an effective molecular formula of $C_{10}H_{10}$ (C_nH_m) which gives the degree of unsaturation as $(n - m/2 + 1) = 10 - 5 + 1 = 6$. The compound contains a combined total of six rings and/or π bonds.

2. IR and ^{13}C NMR spectra establish the presence of a ketone (^{13}C resonance at 191.9 ppm).

3. The coupling pattern in the aromatic region of the ^1H NMR spectrum shows that the compound must be a "1,2,4-trisubstituted benzene". The proton at 7.65 ppm (H_5) clearly has no *ortho* couplings and must be the isolated proton in the spin system sandwiched between two substituents. The two protons at 7.25 and 6.85 ppm each have a large *ortho* coupling so they must be adjacent to each other. The proton at 7.25 ppm (H_7) has an additional *meta* coupling so it must be the proton meta to the proton at 7.65 ppm. The remaining proton at 6.85 ppm (H_8) must be the proton *para* to the proton at 7.65 ppm.

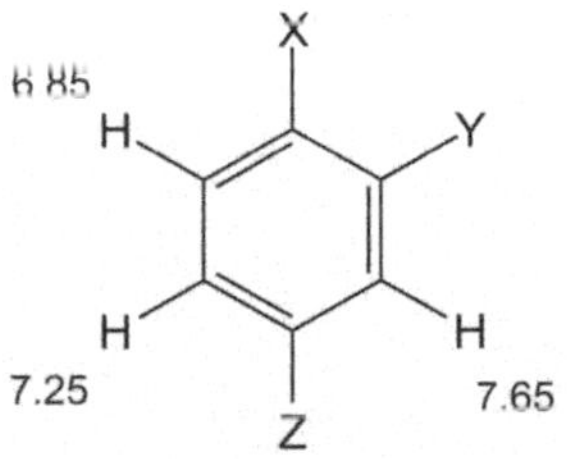

4. There are also signals in the ^1H NMR spectrum for an OCH$_2$ group at 4.48 ppm (H_2), a CH$_2$ group at 2.76 ppm (H_3) and a CH$_3$ group at 2.28 ppm (H_{11}). The CH$_2$ groups are spin-coupled affording the OCH$_2$CH$_2$ fragment.

^{1}H NMR spectrum of 6-methyl-4-chromanone (CDCl$_3$, 500 MHz)

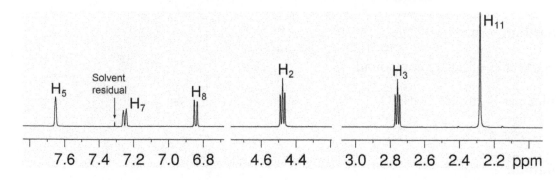

^{13}C{^{1}H} NMR spectrum of 6-methyl-4-chromanone (CDCl$_3$, 125 MHz)

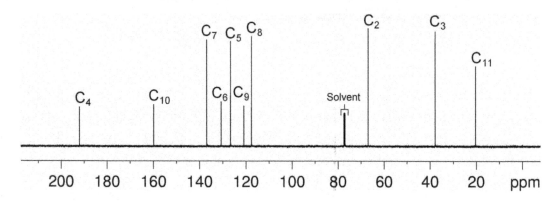

5. The me-HSQC spectrum easily identifies the protonated carbons. The isolated aromatic proton at 7.65 ppm correlates to the carbon at 126.7 ppm (C_5). The aromatic proton at 7.25 ppm correlates to the carbon at 137.0 ppm (C_7). The aromatic proton at 6.85 correlates to the carbon at 117.6 ppm (C_8).

6. In the me-HSQC spectrum, the OCH_2 carbon is at 67.0 ppm (C_2), the CH_2 carbon is at 37.8 ppm (C_3) and the CH_3 carbon is at 20.4 ppm (C_{11}).

7. The three aromatic carbons which bear the substituents appear at 159.9, 130.7 and 120.9 ppm. The low-field carbon resonance (159.9 ppm) must be the carbon bearing an oxygen atom based on its chemical shift.

8. The aromatic ring and ketone functionality accounts for five degrees of unsaturation. There are no signals for an additional double bond thus there must be an additional ring in the molecule.

^{1}H–^{13}C me-HSQC spectrum of 6-methyl-4-chromanone (CDCl$_3$, 500 MHz)

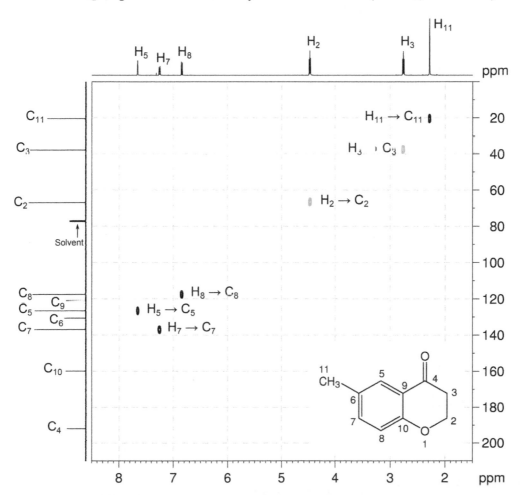

9. In the HMBC spectrum, the correlations from the aromatic protons at 7.65 and 7.25 ppm to the methyl carbon at 20.4 ppm (H_5–C_{11} and H_7–C_{11}) place the CH_3 group as a substituent between these aromatic protons on the aromatic ring (*i.e.* position Z).

10. The methylene to ketone correlations in the HMBC spectrum (H_2–C_4 and H_3–C_4) afford the $OCH_2CH_2C{=}O$ fragment.

11. The H_2–C_{10} correlation in the HMBC spectrum indicates that the OCH_2 group is also a substituent of the aromatic ring.

12. Note that the H_3–C_3 and H_{11}–C_{11} correlations are visible in the HMBC spectrum as large doublets.

^1H–^{13}C HMBC spectrum of 6-methyl-4-chromanone (CDCl$_3$, 500 MHz)

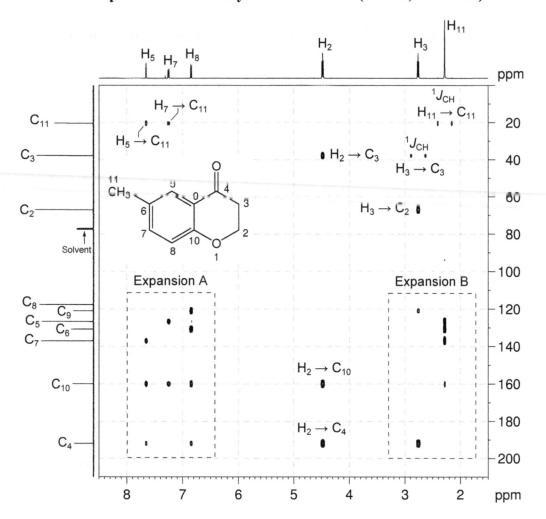

13. The aromatic proton to ketone carbon correlations (H_5–C_4 and H_8–C_4) in the HMBC spectrum (expansion A) indicate that the ketone group is bound to the benzene ring.

^1H–^{13}C HMBC spectrum of 6-methyl-4-chromanone – expansion A

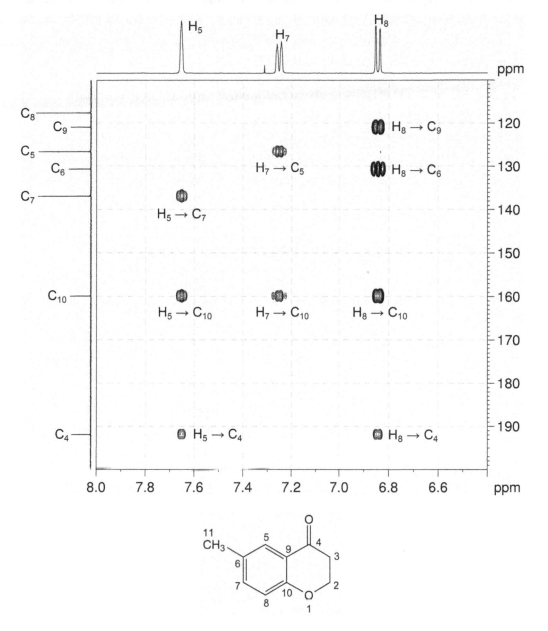

14. There are two possible isomers which are not easily distinguished by the HMBC spectrum. In Isomer A, H_5–C_4 is a three-bond correlation while H_8–C_4 is a four-bond correlation. In Isomer B, H_5–C_4 is now a four bond correlation while H_8–C_4 is a three-bond correlation.

A   B

15. In the HMBC spectrum (expansion A), H_5 correlates to the protonated aromatic carbon signal at 137.0 ppm (C_7) and the quaternary aromatic carbon signal at 159.9 ppm. H_7 correlates to the protonated aromatic carbon signal at 126.7 ppm (C_5) and the quaternary aromatic carbon signal at 159.9 ppm. As both H_5 and H_7 correlate to the signal at 159.9 ppm, this signal may be assigned to the carbon *meta* to H_5 and H_7 (C_{10}). The downfield shift of C_{10} is consistent with the fact that C_{10} is bound to an O atom, pointing towards Isomer A as the correct structure.

16. In expansion B of the HMBC spectrum, the methyl protons (H_{11}) correlate to protonated aromatic carbon signals at 126.7 and 137.0 (C_5 and C_7, respectively) and to quaternary aromatic carbon signal at 130.7 ppm which can be assigned to C_6. The H_{11} correlation to the signal at 159.9 ppm (previously assigned to C_{10}) is a five-bond coupling and again makes it difficult to distinguish between Isomers A and B.

^{1}H–^{13}C HMBC spectrum of 6-methyl-4-chromanone – expansion B

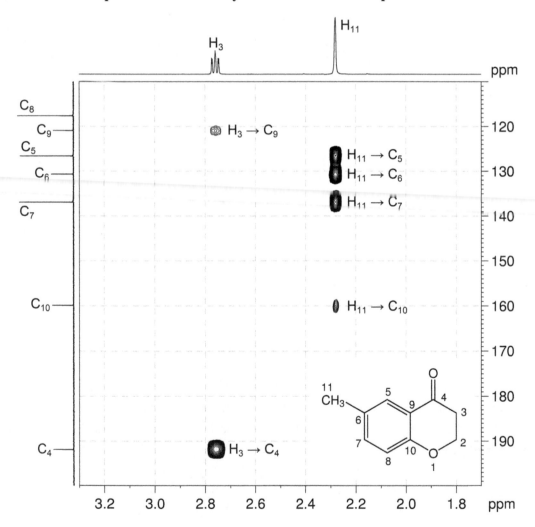

18. The correct structure of the compound can be determined using the INADEQUATE spectrum which shows direct $^{13}C-^{13}C$ connectivity.

19. The methyl carbon C_{11} has a correlation to quaternary aromatic carbon C_6.

20. C_6 has correlations to protonated aromatic carbons C_5 and C_7.

21. C_5 has a correlation to quaternary aromatic carbon C_9 which in turn has correlations to quaternary carbon C_{10} and ketone carbon C_4. These correlations afford the fragment below which confirms **Isomer A** as the correct compound.

INADEQUATE spectrum of 6-methyl-4-chromanone (CDCl₃, 150 MHz)

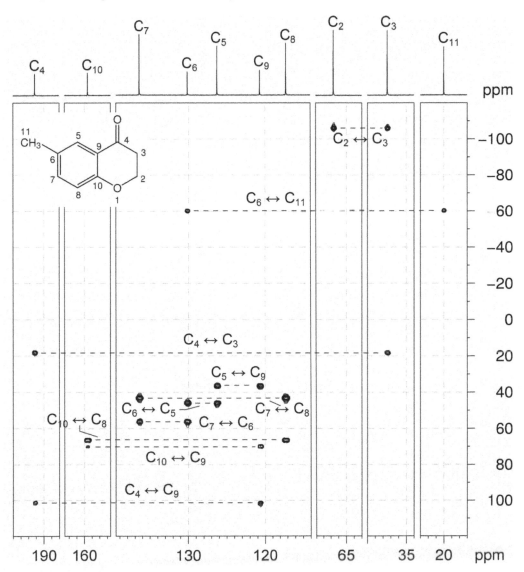

Problem 62

Question:

Identify the following compound.

Molecular Formula: $C_{10}H_{18}O$

IR: 1727 (s) cm^{-1}

Solution:

Citronellal

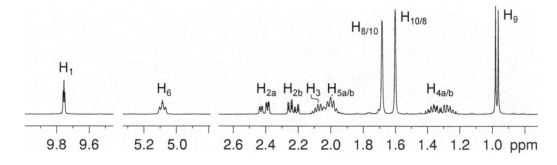

1. The molecular formula is $C_{10}H_{18}O$. Calculate the degree of unsaturation from the molecular formula: ignore the O atoms to give an effective molecular formula of $C_{10}H_{18}$ (C_nH_m) which gives the degree of unsaturation as $(n - m/2 + 1) = 10 - 9 + 1 = 2$. The compound contains a combined total of two rings and/or π bonds.

2. The $^{13}C\{^{1}H\}$ NMR spectrum has a signal at 203.0 ppm consistent with a carbonyl group. The low-field ^{1}H NMR signal at 9.75 ppm correlates to the low-field carbon at 203.0 ppm in the HSQC spectrum indicating the presence of an aldehyde group. This accounts for one of the degrees of unsaturation and the oxygen in the molecular formula.

3. The ^{1}H NMR signal at 5.08 ppm, and the $^{13}C\{^{1}H\}$ NMR signals at 131.8 and 124.0 ppm are consistent with the presence of a trisubstituted alkene. This accounts for the second degree of unsaturation. The compound contains no additional rings or multiple bonds.

^{1}H NMR spectrum of citronellal (CDCl$_3$, 400 MHz)

$^{13}C\{^1H\}$ NMR spectrum of citronellal (CDCl$_3$, 100 MHz)

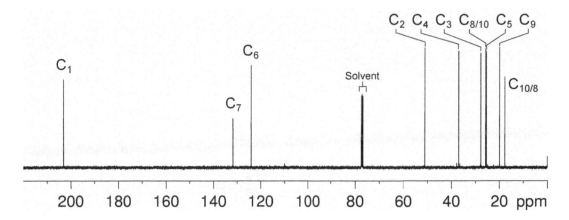

4. The ^1H–^{13}C me-HSQC spectrum shows one quaternary alkene carbon, 3 × CH groups (including aldehyde and alkene protons), 3 × CH$_2$ groups and 3 × CH$_3$ groups. All three CH$_2$ groups are diastereotopic, *i.e.* two chemically non-equivalent protons attached to the same carbon. The molecule must contain a chiral centre.

^1H–^{13}C me-HSQC spectrum of citronellal (CDCl$_3$, 400 MHz)

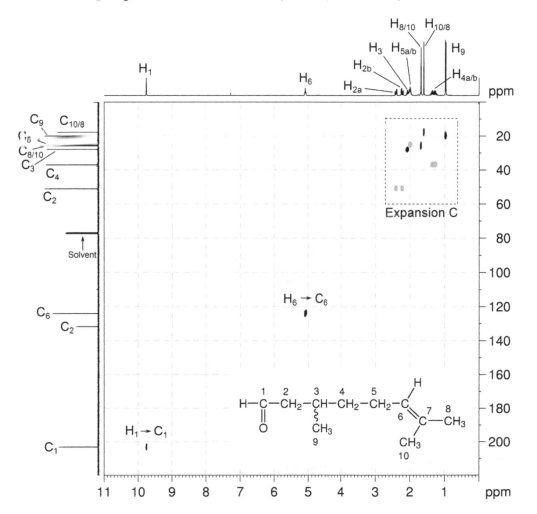

^{1}H–^{13}C me-HSQC spectrum of citronellal – expansion C

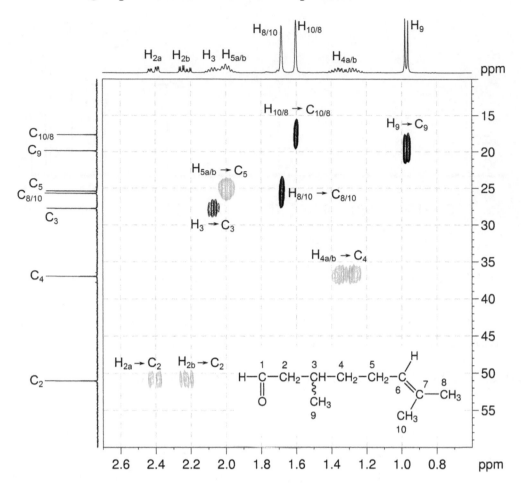

5. The ^{1}H–^{13}C me-HSQC spectrum may be used to identify the signals corresponding to pairs of diastereotopic methylene protons, as pairs of diastereotopic protons correlate to the same ^{13}C resonance. One pair of diastereotopic protons resonates at 2.41 and 2.23 ppm (and both correlate to the ^{13}C resonance at 51.0 ppm), a second pair of diastereotopic protons resonates at 1.36 and 1.28 ppm (and both correlate to the ^{13}C resonance at 36.9 ppm). The final pair of diastereotopic protons resonates at 2.00 ppm (and both correlate to the ^{13}C resonance at 25.4 ppm), and both of these protons are nearly superimposed on the resonance corresponding to an alkyl CH proton at 2.07 ppm.

6. In the ¹H–¹H COSY spectrum (expansion B), the aldehyde proton (H$_1$, 9.75 ppm) couples to the diastereotopic methylene protons at 2.41 and 2.23 ppm. These protons are identified as H$_{2a}$ and H$_{2b}$, and show that a HC(=O)–CH$_2$– fragment is present in the molecule.

¹H–¹H COSY spectrum of citronellal – expansion B (CDCl₃, 400 MHz)

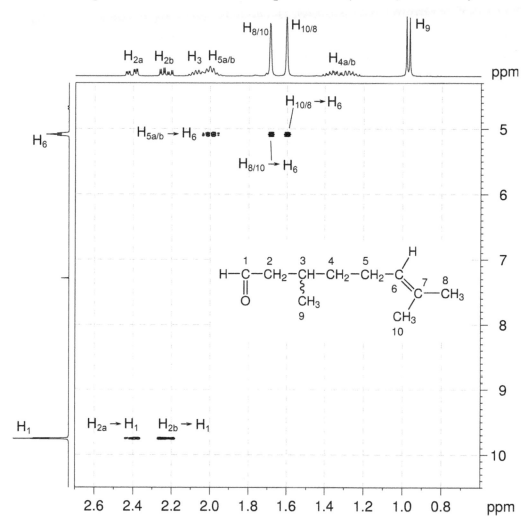

7. Both H_{2a} and H_{2b} correlate to the alkyl CH group at 2.07 ppm in the COSY spectrum (expansion A). This identifies H_3 and extends the fragment to $HC(=O)–CH_2–CH–$.

8. In the COSY spectrum (expansion A), H_3 correlates strongly to the methyl group at 0.97 ppm (identifying H_9), and weakly to the signal at 1.36 ppm (which belong to one proton of the pair of diastereotopic protons at 1.36 and 1.28 ppm – $H_{4a/b}$). The fragment may then be extended further to give $HC(=O)–CH_2–CH(CH_3)–CH_2–$.

9. In the COSY spectrum (expansion A), H_{4a} and H_{4b} correlate strongly each other as well as to the diastereotopic protons at 2.00 ppm (identifying $H_{5a/b}$). The fragment may be extended further to give $HC(=O)–CH_2–CH(CH_3)–CH_2–CH_2–$.

$^1H–^1H$ COSY spectrum of citronellal expansion A

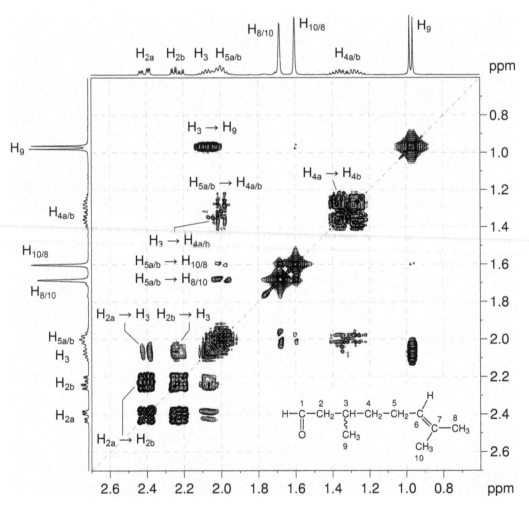

10. In the COSY spectrum (expansion B), $H_{5a/b}$ correlate strongly to the alkene proton at 5.08 ppm (identifying H_6). The molecule fragment may be extended further to give $HC(=O)–CH_2–CH(CH_3)–CH_2–CH_2–CH=$.

11. From the me-HSQC spectrum we know that there is one quaternary and one protonated alkene carbon, so we can extend the molecule fragment to $HC(=O)-CH_2-CH(CH_3)-CH_2-CH_2-CH=C<$.

12. In the COSY spectrum (expansion B), H_6 correlates to both remaining CH_3 groups (at 1.68 and 1.60 ppm). The structure of the molecule must therefore be $HC(=O)-CH_2-CH(CH_3)-CH_2-CH_2-CH=C(CH_3)_2$ – **citronellal**. It is not possible to conclusively identify which methyl group corresponds to H_8 and which is H_{10}.

13. Expansion A of the COSY spectrum contains two long-range $^5J_{H-H}$ correlations between $H_{5a/5b}$ and each of the geminal methyl groups.

14. A check on the correlations in the HMBC spectrum shows that all of the observed correlations are reasonable two- and three-bond correlations.

$^1H-^{13}C$ HMBC spectrum of citronellal – expansion D (CDCl$_3$, 400 MHz)

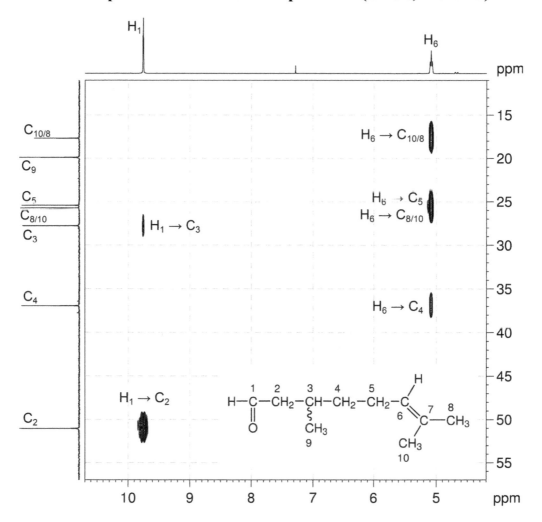

^1H–^{13}C HMBC spectrum of citronellal – expansion E

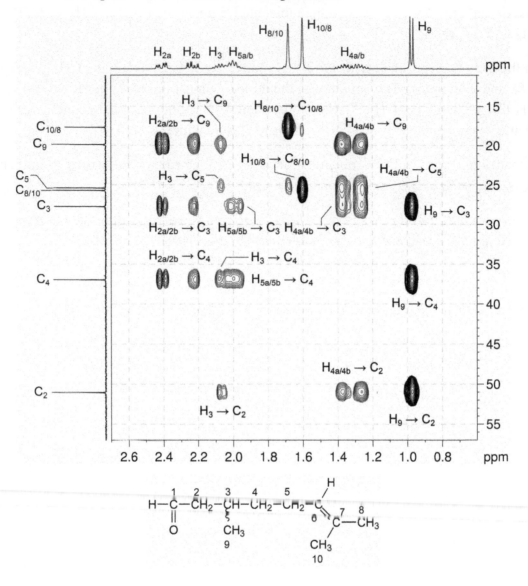

^1H–^{13}C HMBC spectrum of citronellal – expansion F

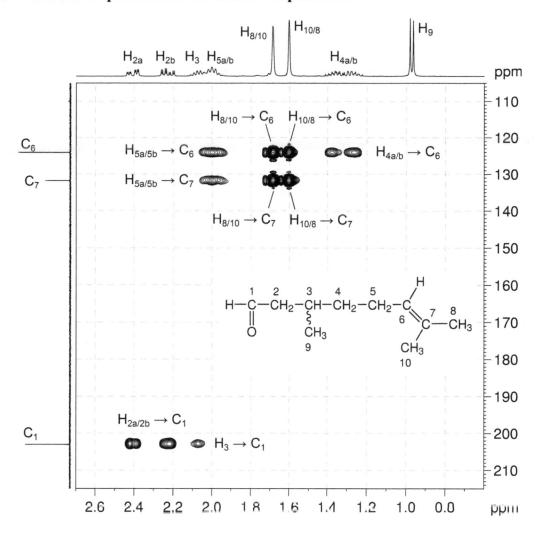

Question:

Identify the following compound, including any relative stereochemistry.

Molecular Formula: $C_7H_8O_2$

IR: 1765 cm^{-1}

Solution:

Oxabicyclo[3.3.0]oct-6-en-3-one

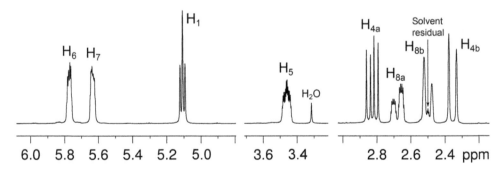

1. The molecular formula is $C_7H_8O_2$. Calculate the degree of unsaturation from the molecular formula: ignore the O atoms to give an effective molecular formula of C_7H_8 (C_nH_m) which gives the degree of unsaturation as $(n - m/2 + 1) = 7 - 4 + 1 = 4$. The compound contains a combined total of four rings and/or π bonds.

2. IR and 1D NMR spectra establish the presence of an ester (^{13}C resonance at 177.0 ppm). The ester group accounts for all oxygen atoms in the molecular formula.

3. In the ^1H NMR spectrum, there are eight separate signals for the eight protons in the molecule.

^1H NMR spectrum of oxabicyclo[3.3.0]oct-6-en-3-one (DMSO-d_6, 400 MHz)

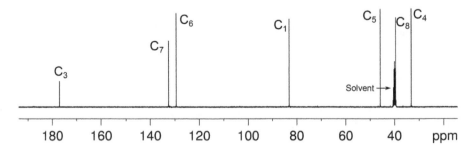

$^{13}C\{^1H\}$ NMR spectrum of oxabicyclo[3.3.0]oct-6-en-3-one (DMSO-d_6, 100 MHz)

4. In the me-HSQC spectrum, there are two resonances in the alkene region at ^{1}H 5.77 / ^{13}C 129.5.ppm (H$_6$ / C$_6$) and ^{1}H 5.63 / ^{13}C 132.5.ppm (H$_7$ / C$_7$).

5. In the me-HSQC spectrum, there are resonances for two CH groups at ^{1}H 5.11 / ^{13}C 83.1.ppm (H$_1$ / C$_1$) and ^{1}H 3.46 / ^{13}C 45.9.ppm (H$_5$ / C$_5$). On the basis of chemical shift, C$_1$ is likely bound to an oxygen atom.

6. In the me-HSQC spectrum, there are also two CH$_2$ groups, both with diastereotopic protons at ^{1}H 2.83, 2.36 / ^{13}C 33.2 ppm (H$_{4a}$, H$_{4b}$ / C$_4$) and ^{1}H 2.68, 2.50 / ^{13}C 39.5 ppm (H$_{8a}$, H$_{8b}$ / C$_8$). The diastereotopic protons indicate there is either a ring or chiral carbon in the molecule.

7. The ester and alkene functional groups account for two degrees of unsaturation. There are two additional degrees on unsaturation and these are likely due to the presence of rings in the molecule.

^{1}H–^{13}C me-HSQC spectrum of oxabicyclo[3.3.0]oct-6-en-3-one (DMSO-d_6, 400 MHz)

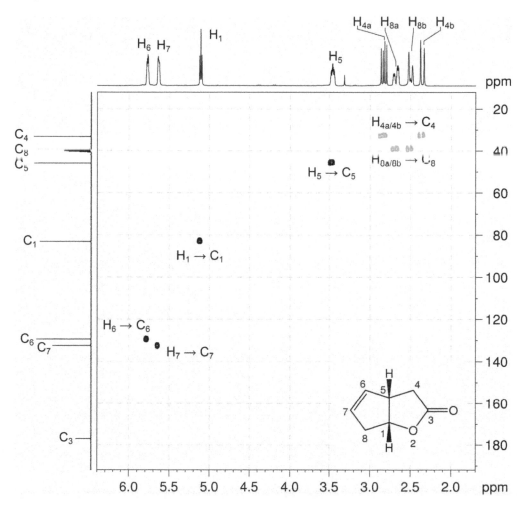

8. In the COSY spectrum, H_6 and H_7 are coupled affording a CH=CH fragment.

9. In the COSY spectrum, H_5 couples to H_6 and H_1 affording a CH=CH–CH–CH fragment.

10. In the COSY spectrum, H_1 further couples to H_{8a} affording the CH=CH–CH–CH–CH$_2$ fragment

11. The H_6–H_{8a}, H_7–H_{8a}, H_6–H_{8b} and H_7–H_{8b} correlations in the COSY spectrum affords the ring structure below:

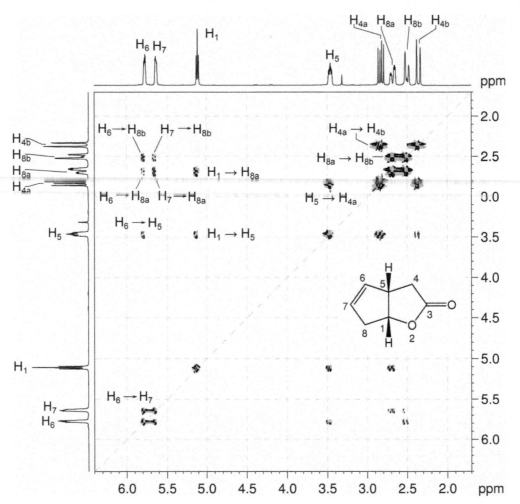

^{1}H–^{1}H COSY spectrum of oxabicyclo[3.3.0]oct-6-en-3-one (DMSO-d_6, 400 MHz)

12. In the COSY spectrum, the H_5–H_{4a} correlation indicates there is a CH_2 group bound to C_5.

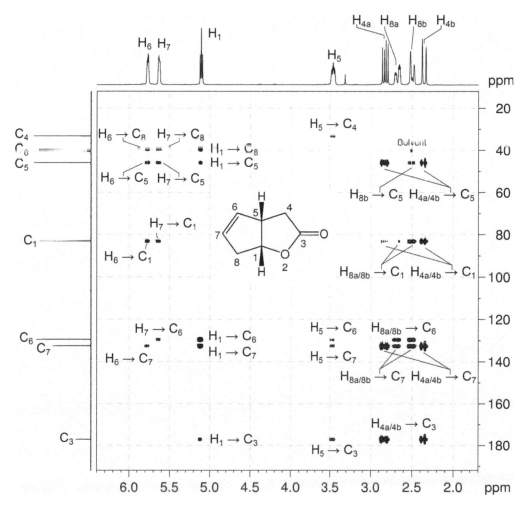

13. We have now accounted for three degrees of unsaturation (ester group, alkene, ring) and all the atoms of the molecule. The remaining degree of unsaturation must be due to a second ring linked up by the ester group.

14. On the basis of chemical shift, C_1 rather than C_4 is bound to the oxygen atom of the ester group.

15. All correlations in the HMBC spectrum are consistent with the structure thus obtained.

^{1}H–^{13}C HMBC spectrum of oxabicyclo[3.3.0]oct-6-en-3-one (DMSO-d_6, 400 MHz)

16. The relative stereochemistry of the molecule can be determined from the NOESY spectrum. In particular, the H_1–H_5 correlation in the NOESY spectrum indicates that both H_1 and H_5 are on the same side of the molecule.

17. Also from the NOESY spectrum, the H_1–H_{8a} and H_5–H_{4a} correlations indicate that the H_{8a} and H_{4a} protons are also on the same side of the molecule as H_1 and H_5.

^1H–^1H **NOESY spectrum of oxabicyclo[3.3.0]oct-6-en-3-one (DMSO-d_6, 400 MHz)**

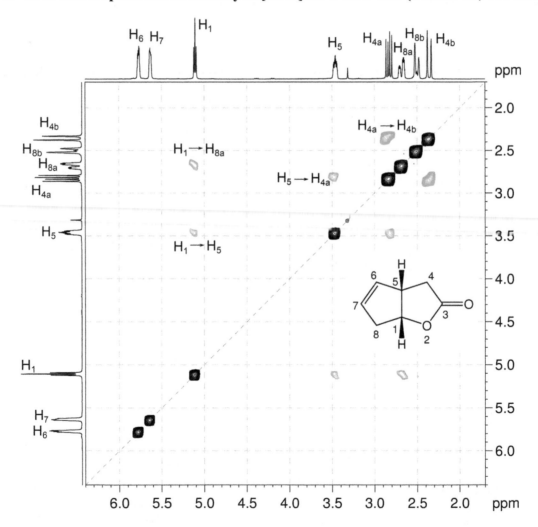

Question:

Identify the following compound.

Molecular Formula: $C_{13}H_{16}N_2O_2$

IR: 3305, 1629, 1620, 1566 cm^{-1}

Solution:

Melatonin

1. The molecular formula is $C_{13}H_{16}N_2O_2$. Calculate the degree of unsaturation from the molecular formula: ignore the O atoms, ignore the N atoms and remove one H atom for each N to give an effective molecular formula of $C_{13}H_{14}$ (C_nH_m) which gives the degree of unsaturation as $(n - m/2 + 1) = 13 - 7 + 1 = 7$. The compound contains a combined total of seven rings and/or π bonds.

2. From the ^1H NMR spectrum, there are two exchangeable protons of integration one – one which readily exchanges, and another which requires more forcing conditions. The latter is likely due to a secondary amide group, while the former could be either a secondary amine or an alcohol group.

3. There are four protons in the aromatic region of the ^1H NMR spectrum, but the coupling pattern does not match that of a disubstituted aromatic ring. This suggests that there may be one aromatic ring which is trisubstituted, and a second unsaturated system somewhere in the molecule which pushes the chemical shift of the proton into the aromatic region.

4. There is a singlet of integration three at 3.78 ppm in the ^1H NMR spectrum. This chemical shift and integration is consistent with the presence of a methoxy ($-OCH_3$) group.

5. With the presence of an amide group and a methoxy group, we have now accounted for all the oxygen atoms in the molecular formula, so the remaining exchangeable proton must be due to a secondary amine group.

6. There are two multiplets of integration two at 3.32 and 2.79 ppm in the ^{1}H NMR spectrum. These are likely due to two methylene groups. The last remaining signal is a singlet of integration three at 1.82 ppm – this must be due to an isolated methyl group.

^{1}H NMR spectrum of melatonin – expansion 1 (DMSO-d_6, 500 MHz)

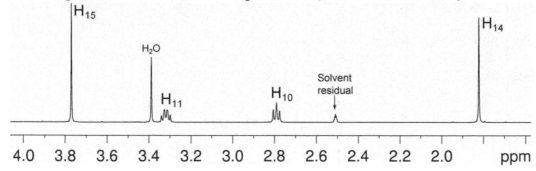

^{1}H NMR spectrum of melatonin – expansion 2

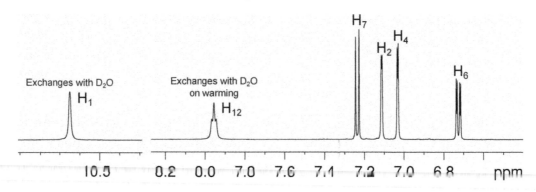

7. From the ^{1}H–^{1}H COSY spectrum, we can see a correlation between the amide proton at 7.95 ppm and the methylene group at 3.32 ppm. This methylene group also correlates to the second methylene group at 2.79 ppm. So we can identify a –$CH_2CH_2NHC(=O)$– fragment.

^{1}H–^{1}H COSY spectrum of melatonin (DMSO-d_6, 500 MHz)

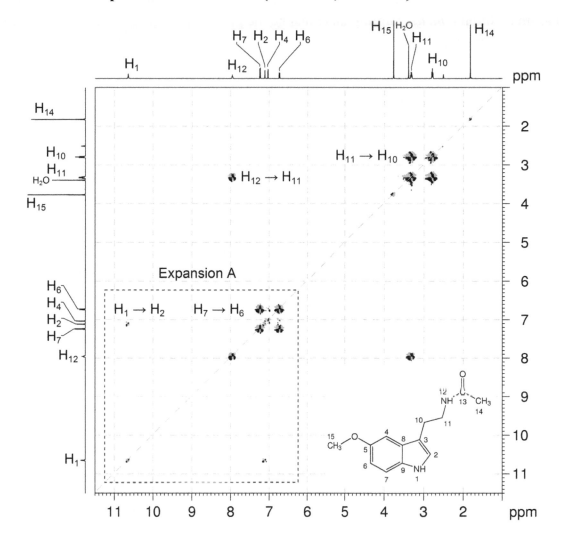

8. There is also a correlation between the amine proton at 10.6 ppm and the "aromatic" signal at 7.11 ppm. We can therefore identify an –NHCH= fragment.

1. The remaining aromatic protons must be due to a trisubstituted
    aromatic ring, and from the coupling pattern we can identify the
    substitution pattern as that of a "1,2,4-trisubstituted benzene".
    The proton at 7.03 ppm (H4) does not have any *ortho* couplings
    and must be the isolated proton sandwiched between two
    substituents. The two protons at 6.72 and 7.23 ppm each have a
    large *ortho* coupling so they must be adjacent to each other. The proton at 6.72 ppm (H6)
    has an additional *meta* coupling and must be the proton *meta* to the proton at 7.03 ppm.
    The remaining proton at 7.23 ppm (H7) must be the proton *para* to the proton at
    7.03 ppm.

2. Summarising the fragments so far identified gives us:

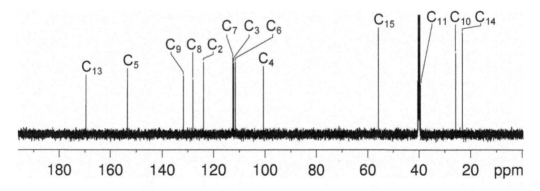

3. These fragments account for all the hydrogen, nitrogen and oxygen atoms, but only 12
    carbon atoms. According to the molecular formula there must be one carbon atom we
    have so far not identified. If we study the $^{13}C\{^1H\}$ NMR spectrum, there are eight signals
    between 100 and 160 ppm, and we have only thus far identified seven. The missing
    carbon atom must be due to a quaternary aromatiC– or alkene-type carbon.

$^{13}C\{^1H\}$ NMR spectrum of melatonin (DMSO-d_6, 125 MHz)

---

4. The ^{1}H–^{13}C me-HSQC spectrum identifies each protonated carbon atom: the isolated methyl group at 23.1 ppm, the two methylene groups at 27.7 and 39.9 ppm (the second signal is hidden in the DMSO-d_6 solvent signal), and the methoxy group at 55.8 ppm. In the aromatic region, the protonated carbons for the trisubstituted ring occur at 100.6, 111.5 and 112.4. The protonated carbon for the amine-bound carbon atom resonates at 123.7 ppm.

^{1}H–^{13}C me-HSQC spectrum of melatonin (DMSO-d_6, 500 MHz)

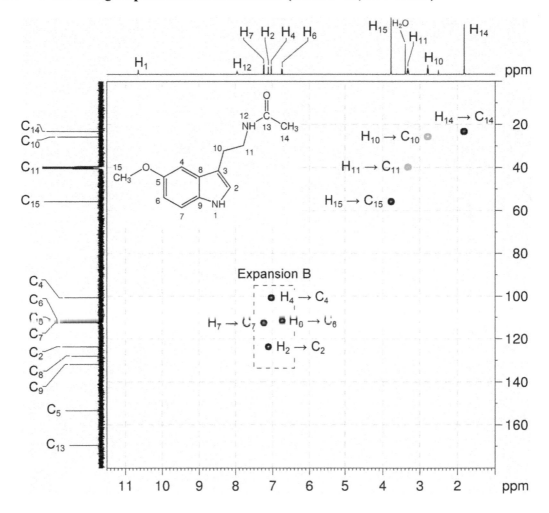

5.   The ^{1}H–^{15}N HSQC spectrum shows one-bond correlations between proton and nitrogen nuclei. The ^{1}H–^{15}N HSQC spectrum identifies the chemical shifts of the two nitrogen nuclei – the amine nitrogen at 128.9 ppm and the amide nitrogen at 120.0 ppm.

^{1}H–^{15}N HSQC spectrum of melatonin (DMSO-d_6, 500 MHz)

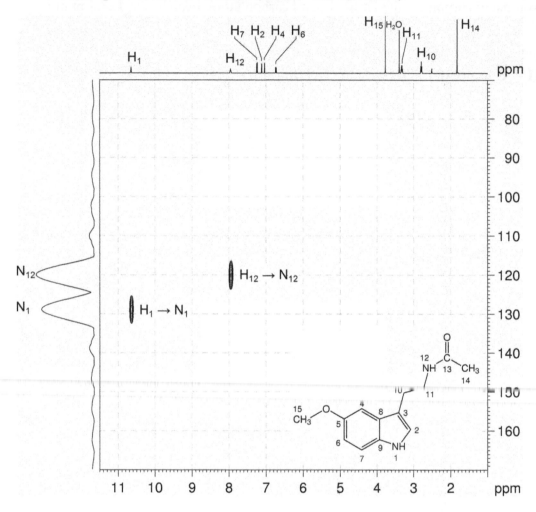

6. The ^{1}H–^{15}N HMBC spectrum identifies protons that are two or three bonds removed from the nitrogen nucleus. The amide nitrogen correlates to one methylene group, and also to the isolated methyl group, so we can combine these two fragments to give a substituted acetamide: –CH$_2$CH$_2$NHC(=O)CH$_3$.

^{1}H–^{15}N HMBC spectrum of melatonin (DMSO-d_6, 500 MHz)

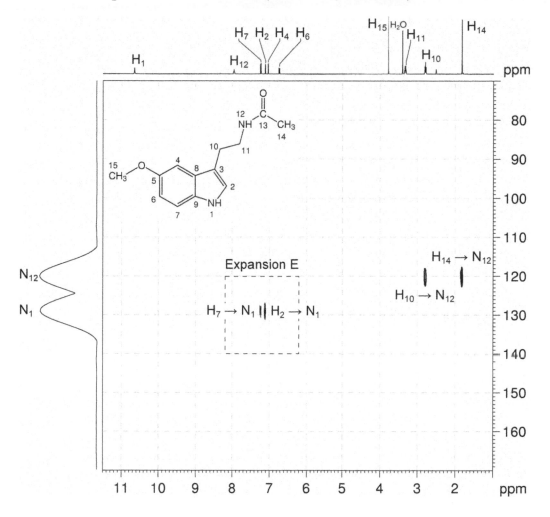

7. The amine nitrogen correlates to the previously identified neighbouring proton at 7.11 ppm (H$_2$), but it also correlates to the aromatic ring signal at 7.23 ppm (H$_7$), meaning the nitrogen must be directly bound to the aromatic ring adjacent to the aromatic proton at 7.23 ppm. Redrawing the known fragments gives:

8. The two alkene carbons must be bound, giving:

$$-CH_2-CH_2-NH-\underset{\underset{O}{\|}}{C}-CH_3 \qquad -O-CH_3$$

9. The alkene must be bound to the aromatic ring to give a fused heterocycle (binding the methoxy group and the amide group to the alkene carbon leaves two unsubstituted aromatic positions that cannot be filled). There are therefore two possible compounds that fit the spectra considered so far:

A   B

10. It is possible to differentiate these using either the NOESY spectrum, or the $^1H-^{13}C$ HMBC spectrum.

11. In the NOESY spectrum (expansion C), there are correlations between the methoxy group (H$_{15}$) and aromatic protons H$_4$ and H$_6$, which places the methoxy group between these two protons on the six-membered aromatic ring, rather than on the five-membered heterocyclic ring. Isomer A (melatonin) is therefore the correct structure.

^{1}H–^{1}H NOESY spectrum of melatonin – expansion C (DMSO-d_6, 500 MHz)

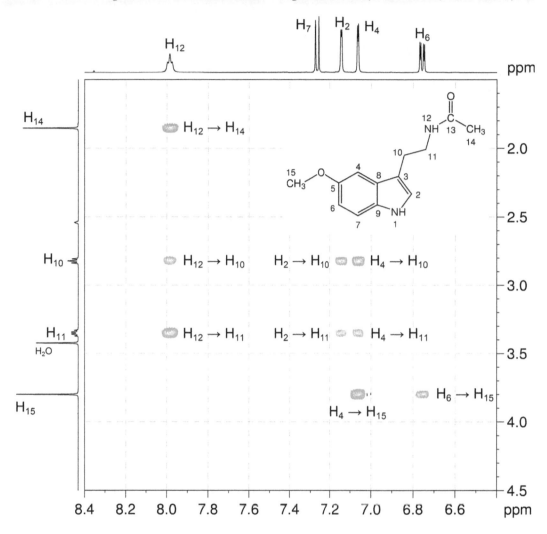

$^1H-^1H$ NOESY spectrum of melatonin – expansion D

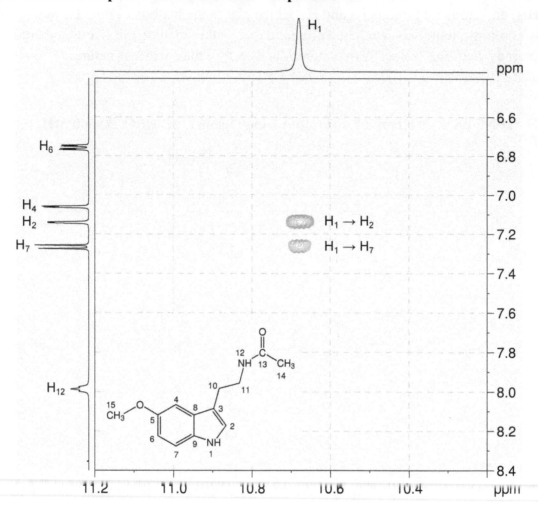

12. In the ^1H–^{13}C HMBC spectrum (expansion J), there is a correlation between the methoxy group and the quaternary aromatic carbon (at 153.4 ppm). The three protons of the six-membered aromatic ring also correlate to this signal (expansion I), again placing the methoxy group on the six-membered ring, rather than the five-membered heterocycle. Once again, Isomer A (melatonin) is identified as the correct structure.

^1H–^{13}C HMBC spectrum of melatonin – expansion F (DMSO-d_6, 500 MHz)

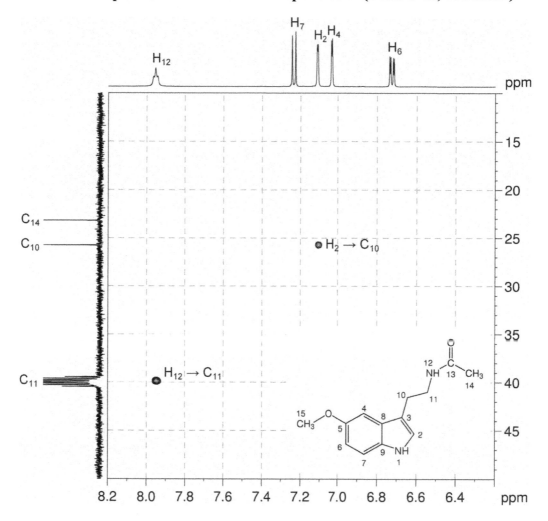

^1H–^{13}C HMBC spectrum of melatonin – expansion G

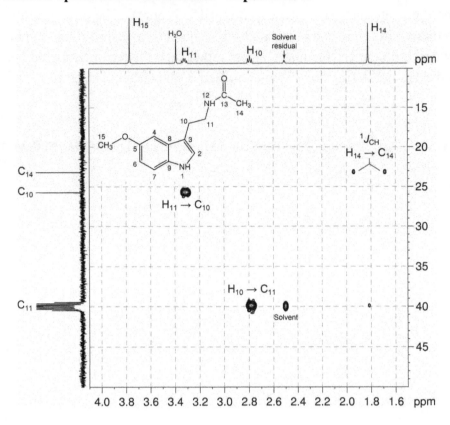

^1H–^{13}C HMBC spectrum of melatonin – expansion H

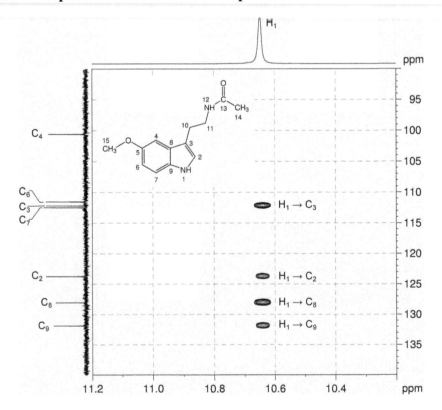

^1H–^{13}C HMBC spectrum of melatonin - expansion I

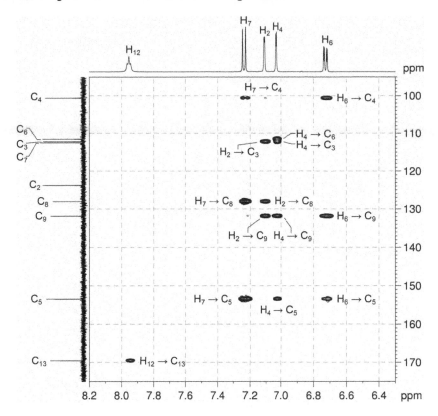

^1H–^{13}C HMBC spectrum of melatonin - expansion J

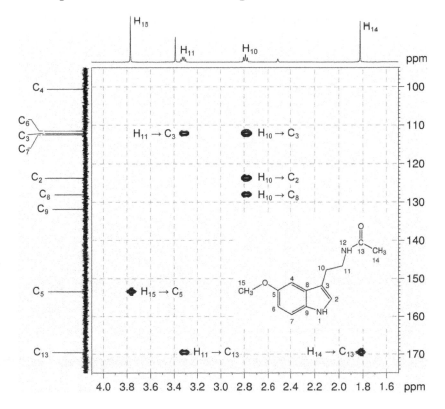

Problem 65

Question:

Identify the following compound.

Molecular Formula: $C_{10}H_{14}O$

IR: 1675 (s), 898 (s) cm^{-1}

Solution:

Carvone

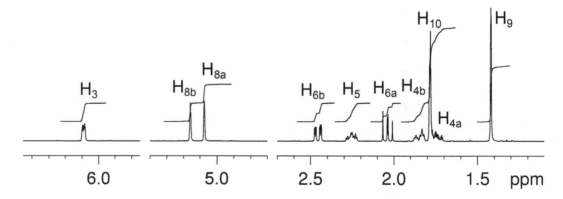

1. The molecular formula is $C_{10}H_{14}O$. Calculate the degree of unsaturation from the molecular formula: ignore the O atom to give an effective molecular formula of $C_{10}H_{14}$ (C_nH_m) which gives the degree of unsaturation as $(n - m/2 + 1) = 10 - 7 + 1 = 4$. The compound contains a combined total of four rings and/or π bonds.

2. IR and $^{13}C\{^1H\}$ spectra establish the presence of a ketone (^{13}C resonance at 197.7 ppm). The ketone group accounts for the single O atom in the molecular formula as well as one degree of unsaturation.

3. In the ^1H NMR spectrum, there are signals for three alkene protons at 6.09, 4.66 and 4.57 ppm (H3, H8b and H8a, respectively).

^1H NMR spectrum of carvone (C$_6$D$_6$, 500 MHz)

$^{13}C\{^1H\}$ NMR spectrum of carvone (C_6D_6, 125 MHz)

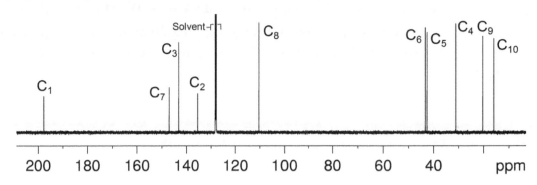

4. In the me-HSQC spectrum, the signals for H_{8a} and H_{8b} correlate to a single carbon signal at 110.4 ppm (C_8) indicating they are both bound to the same carbon atom. H_3 correlates to the carbon signal at 143.2 ppm (C_3).

^1H–^{13}C me-HSQC spectrum of carvone (C_6D_6, 500 MHz)

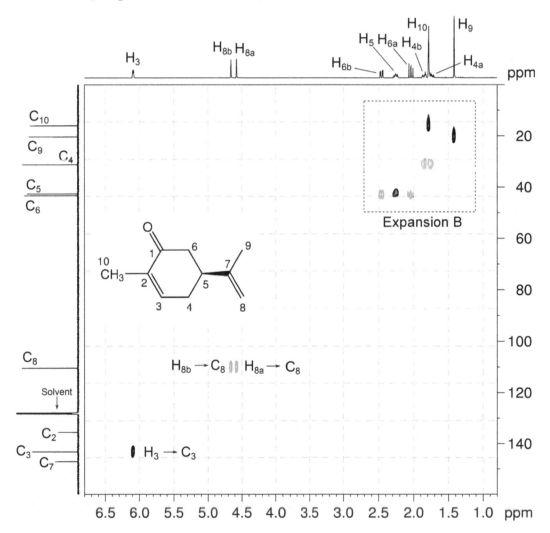

5. In the me-HSQC spectrum (expansion B), there are also signals for two CH$_2$ groups with diastereotopic protons at ^{1}H 2.46, 2.04 ppm / ^{13}C 43.3 ppm (H$_{6a}$,H$_{6b}$/C$_6$) and ^{1}H 1.85, 1.73 ppm / ^{13}C 31.2 ppm (H$_{4a}$,H$_{4b}$/C$_4$), one CH group at ^{1}H 2.26 ppm / ^{13}C 42.6 ppm (H$_5$/C$_5$) and two methyl groups at ^{1}H 1.78 ppm / ^{13}C 16.0 ppm (H$_{10}$/C$_{10}$) and ^{1}H 1.42 ppm / ^{13}C 20.3 ppm (H$_9$/C$_9$).

^{1}H–^{13}C me-HSQC spectrum of carvone – expansion B

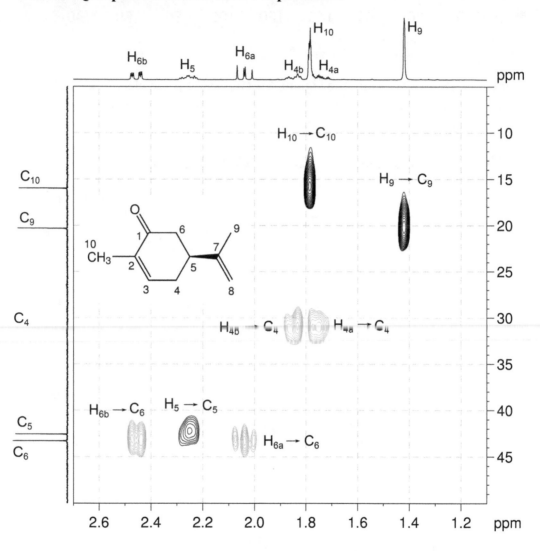

6. In the COSY spectrum, the H$_3$–H$_{4a}$/H$_{4b}$, H$_{4a}$/H$_{4b}$–H$_5$ and H$_5$–H$_{6a}$/H$_{6b}$ correlations afford the =CH–CH$_2$–CH–CH$_2$ fragment.

¹H–¹H COSY spectrum of carvone (C₆D₆, 500 MHz)

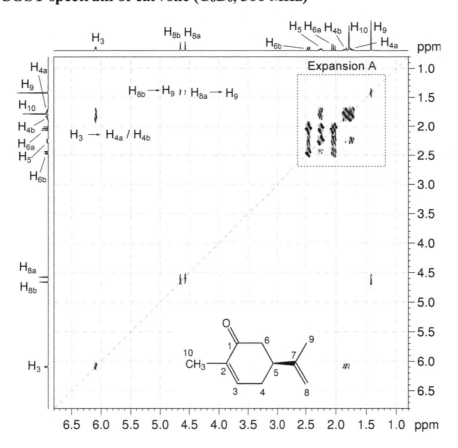

¹H–¹H COSY spectrum of carvone – expansion A

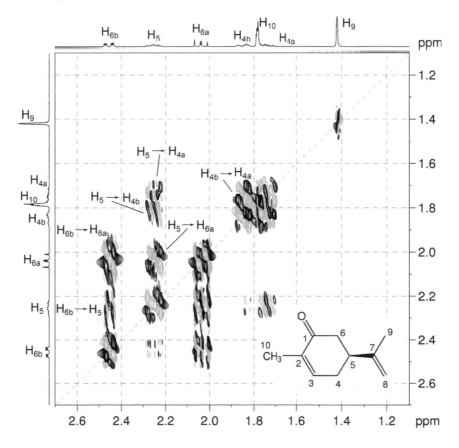

7. In the HMBC spectrum, there are correlations from alkene proton H_3, methylene protons H_{6a}/H_{6b}, methine proton H_5 and methyl protons H_{10} to the ketone carbon at 197.7 ppm (C_1). There are no correlations from methylene protons H_{4a}/H_{4b} to the ketone group. There must be a ring which contains the ketone group to account for these correlations.

$^1H–^{13}C$ HMBC spectrum of carvone (C_6D_6, 500 MHz)

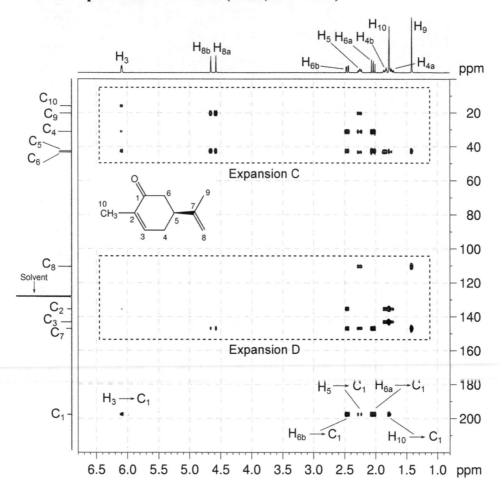

8. In the HMBC spectrum (expansion D), the correlations from H_{6b}, H_{4b}, H_{10} and H_{4a} to the quaternary carbon signal at 135.5 ppm identifies the carbon atom completing the ring (C_2). The correlation from methyl protons H_{10} to alkene carbon C_3 and the reciprocal correlation from H_3 to C_{10} indicates the methyl group is part of the ring.

9. There are two possible ways of assembling the ring structure – one to form a five-membered ring and the other to form a six-membered ring.

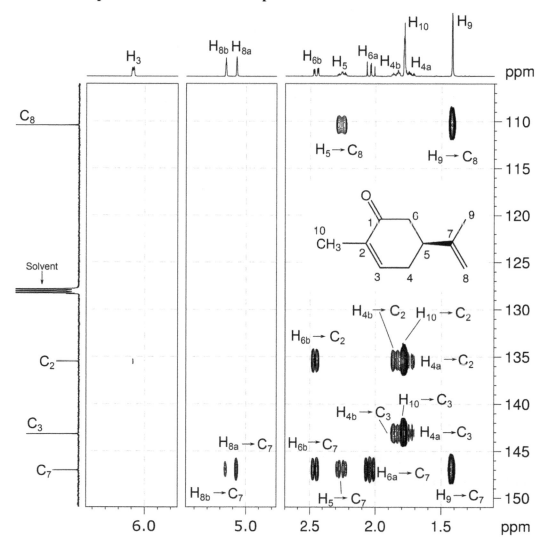

^1H–^{13}C HMBC spectrum of carvone – expansion D

10. In expansions C and D of the HMBC spectrum, the alkene proton signals at 4.66 and 4.57 ppm (H_{8a} and H_{8b}) correlate to the methyl carbon signal at 20.3 ppm (C_9) and quaternary carbon signal at 147.1 ppm (C_7) to afford the $CH_3-C=CH_2$ fragment.

$^1H-^{13}C$ HMBC spectrum of carvone – expansion C

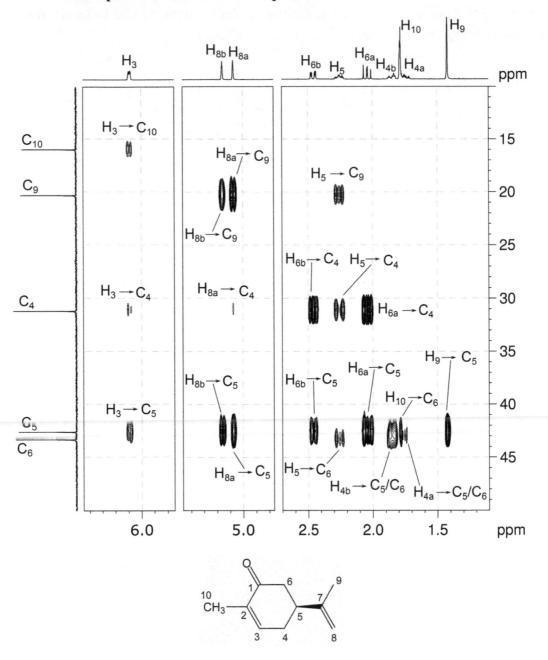

11. The $CH_3-C=CH_2$ fragment can be attached to either of the two possible rings to give two possible structures:

12. The H_{8a}–C_5, H_{8b}–C_5, H_5–C_9 and H_9–C_5 correlations would be four-bond correlations in the structure with the five-membered ring and quite unlikely. These correlations would be three-bond correlations in the structure with the six-membered ring. Thus the correct structure for the molecule is the one with the six-membered ring.

13. While the compound does contain a chiral centre, NMR cannot establish the absolute stereochemistry of the chiral centre.

Question:

The ^{1}H and ^{13}C{^{1}H} NMR spectra of haloperidol ($C_{21}H_{23}ClFNO_2$) recorded in Acetone-d_6 solution at 298 K and 500 MHz are given below.

The ^{1}H NMR spectrum has signals at δ 1.57, 1.81, 1.95, 2.43, 2.44, 2.67, 3.03, 3.86, 7.29, 7.31, 7.43 and 8.14 ppm.

The ^{13}C{^{1}H} NMR spectrum has signals at δ 22.3, 35.6, 38.4, 49.2, 57.6, 70.2, 115.3, 126.6, 127.8, 130.8, 131.4, 134.5, 149.1, 165.4 and 197.8 ppm.

Use the spectra below to assign each proton and carbon resonance.

Solution:

Haloperidol

Proton	Chemical Shift (ppm)	Carbon	Chemical Shift (ppm)
		C_1	165.4
H_2	7.29	C_2	115.3
H_3	8.14	C_3	130.8
		C_4	134.5
		C_5	197.8
H_6	3.03	C_6	35.6
H_7	1.95	C_7	22.3
H_8	2.44	C_8	57.6
H_9	2.43/2.67	C_9	49.2
H_{10}	1.57/1.81	C_{10}	38.4
		C_{11}	70.2
		C_{12}	149.1
H_{13}	7.43	C_{13}	126.6
H_{14}	7.31	C_{14}	127.8
		C_{15}	131.4
OH	3.86		

1. The OH chemical shift can be easily be determined by the exchangeable resonance in the ^{1}H NMR spectrum.

2. The chemical shift of the ketone carbon C$_5$ can also be determined on the basis of chemical shift.

3. Close inspection of the ^{1}H and ^{1}H{^{19}F} NMR spectra show that the aromatic proton signal at 8.14 ppm couples to fluorine and must be either H$_2$ or H$_3$. The signal at 7.43 ppm does not couple to fluorine and must be either H$_{13}$ or H$_{14}$.

4. The chemical shift for H$_7$ can also be assigned to the signal at 1.95 ppm in the ^{1}H NMR spectrum on the basis of coupling pattern (the signal is a quintet and must be coupled to four neighbouring protons).

^{1}H NMR spectrum of haloperidol (acetone-d_6, 500 MHz)

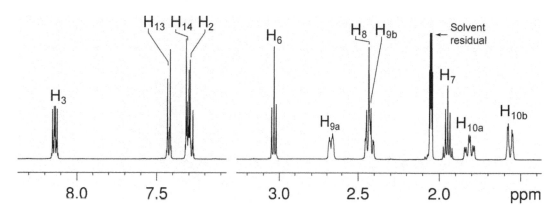

^{1}H{^{19}F} NMR spectrum of haloperidol – expansion (Acetone-d_6, 300 MHz)

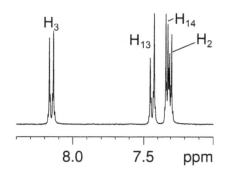

$^{13}C\{^1H\}$ NMR spectrum of haloperidol (acetone-d_6, 125 MHz)

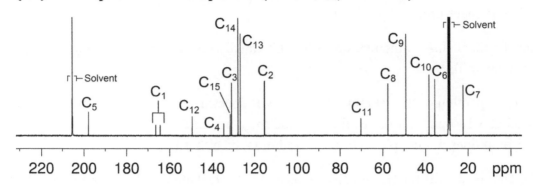

^1H–^1H COSY spectrum of haloperidol – expansion A (acetone-d_6, 500 MHz)

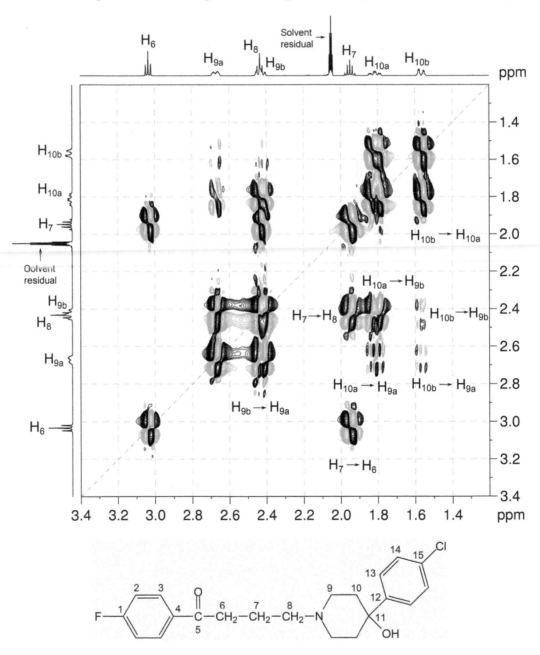

5. In expansion B of the COSY spectrum, the signal at 8.14 ppm couples to the signal at 7.29 ppm (H_2 or H_3) whilst the signal at 7.43 ppm couples to the signal at 7.31 ppm (H_{13} or H_{14}).

^{1}H–^{1}H COSY spectrum of haloperidol – expansion B

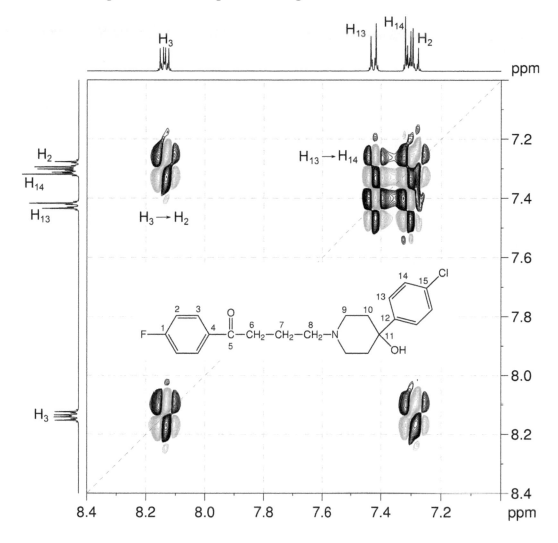

6. The chemical shift for C_7 can be assigned to the signal at 22.3 ppm based on the H_7–C_7 correlation in the me-HSQC spectrum.

7. In expansion C of the me-HSQC spectrum, the diastereotopic CH_2 protons H_{9a}/H_{9b} and H_{10a}/H_{10b} are identified as belonging to the piperidinol ring.

^1H–^{13}C me-HSQC spectrum of haloperidol – expansion C (acetone-d_6, 500 MHz)

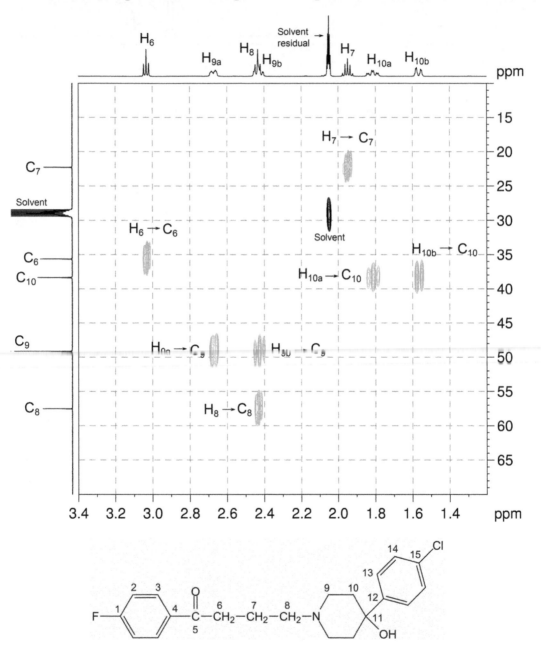

8. In expansion D of the me-HSQC spectrum, the protonated carbons C_2 and C_3 can be assigned to the signals at 115.3 and 130.8 ppm, respectively, as C_2 has a larger coupling to fluorine ($^2J_{C-F} = 22$ Hz) than C_3 ($^3J_{C-F} = 9$ Hz). Thus the proton signals at 7.29 and 8.14 ppm can be assigned to H_2 and H_3, respectively.

^{1}H–^{13}C me-HSQC spectrum of haloperidol – expansion D

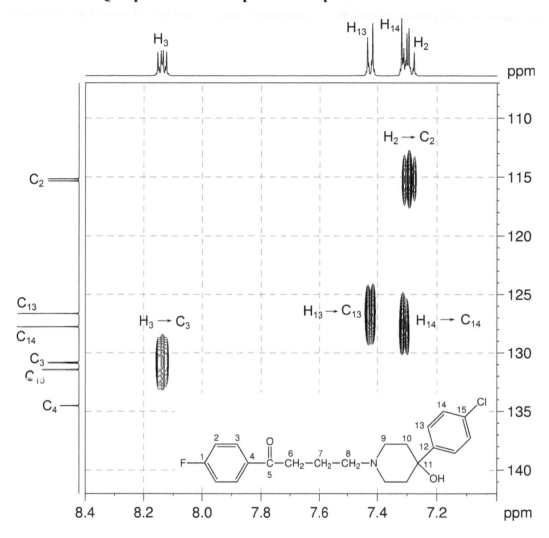

9. C_{11} is the only aliphatic quaternary carbon and can be assigned to the signal at 70.2 ppm which does not have a correlation in the me-HSQC spectrum.

10. In the HMBC spectrum, the H$_{13}$–C$_{11}$ correlation identifies the proton signal at 7.43 ppm as that belonging to H$_{13}$. The signal at 126.6 ppm can now be identified as belonging to C$_{13}$ from the me-HSQC spectrum.

11. The H$_3$–C$_5$ (ketone) correlation in the HMBC spectrum confirms the signal at 8.14 ppm belongs to H$_3$.

12. The H$_6$–C$_5$ (ketone) correlation in the HMBC spectrum identifies the proton signal at 3.03 ppm as that belonging to H$_6$. C$_6$ can now be assigned to the signal at 35.6 ppm using the me-HSQC spectrum.

^{1}H–^{13}C HMBC spectrum of haloperidol (acetone-d_6, 500 MHz)

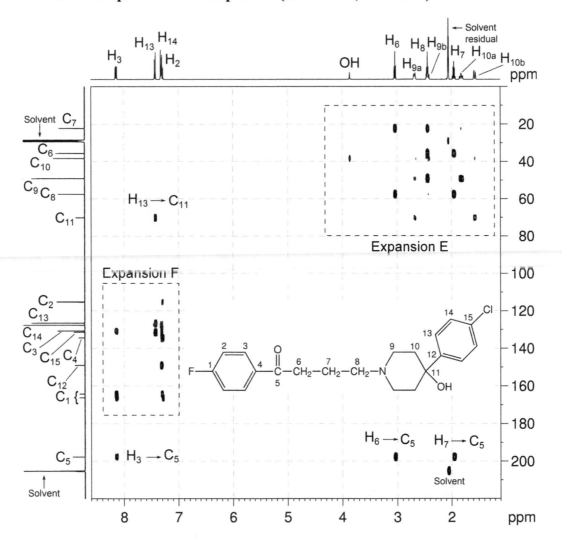

13. The ^{1}H signal at 2.44 ppm belongs to H_8 from the H_8–C_7 and H_8–C_6 correlations in the HMBC spectrum. C_8 can now be assigned to the signal at 57.6 ppm from the me-HSQC spectrum. The H_6–C_8 and H_7–C_8 correlations in the HMBC spectrum confirm the assignment for C_8.

14. The H_8–C_9 correlation in the HMBC spectrum identifies that the signal at 49.2 ppm belongs to C_9. Thus H_9 can be assigned to the signals at 2.43 and 2.67 ppm using the me-HSQC spectrum.

15. The remaining diastereotopic proton signals at 1.57 and 1.81 ppm must be due to H_{10}. Thus the signal at 38.4 ppm is assigned to C_{10} using the me-HSQC spectrum. These assignments can be confirmed by the OH–C_{10} and H_{10a}–C_9 correlations in the HMBC spectrum.

^{1}H–^{13}C HMBC spectrum of haloperidol – expansion E

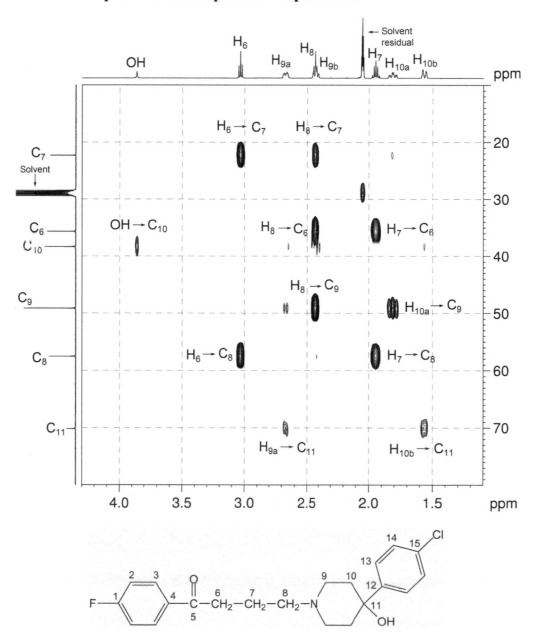

16. Aromatic quaternary carbon C_1 can be assigned to the signal at 165.4 ppm based on the H_3–C_1 and H_2–C_1 correlations in the HMBC spectrum. Note that the signal is a doublet with a large $^1J_{C-F}$ coupling of 251 Hz.

17. C_4 can be assigned to the signal at 134.5 ppm based on the H_2–C_4 correlation in the HMBC spectrum. This signal has a small $^4J_{C-F}$ coupling of 3 Hz just visible in the expansion of the $^{13}C\{^1H\}$ spectrum.

18. C_{15} is assigned to the signal at 131.4 ppm based on the H_{13}–C_{15} and H_{14}–C_{15} correlations in the HMBC spectrum.

19. Finally, C_{12} is assigned to the signal at 149.1 ppm based on the H_{14}–C_{12} correlation in the HMBC spectrum.

^1H–^{13}C HMBC spectrum of haloperidol – expansion F

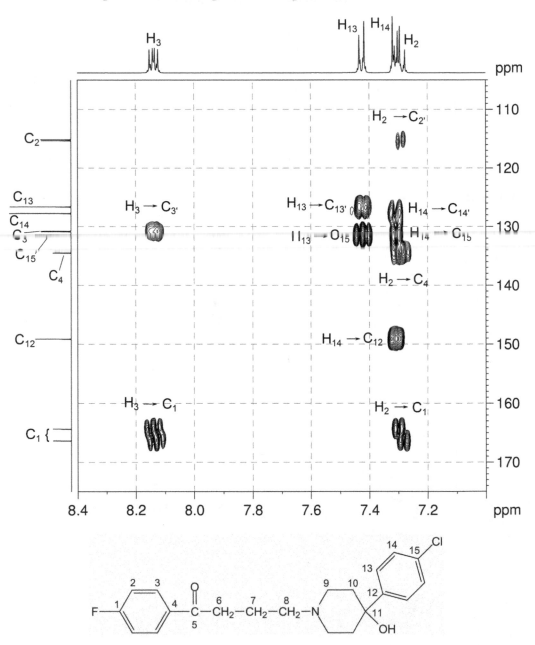

20. Note that there are H_2–$C_{2'}$, H_3–$C_{3'}$, H_{13}–$C_{13'}$ and H_{14}–$C_{14'}$ correlations in the HMBC spectrum which appear to be one-bond correlations but actually arise from the $^3J_{C-H}$ interaction of the proton with the carbon which is *meta* to it.